Cowboys and Cattle Kings

LIFE ON THE RANGE TODAY

COWBOYS
and
CATTLE KINGS
Life on the Range Today

By C. L. SONNICHSEN

NORMAN : UNIVERSITY OF OKLAHOMA PRESS

By C. L. SONNICHSEN

Billy King's Tombstone (Caldwell, Idaho, 1942)
Roy Bean, Law West of the Pecos (New York, 1943)
Cowboys and Cattle Kings: Life on the Range Today
(Norman, 1950)

To H. M. Sonnichsen, *my father*
and Philip Sonnichsen, *my son*

They rode the river with me

Contents

Illustrations

"I Two Faces West

"I 've tried my hand at a lot of other things," Audrey Jane wrote, "—clerking, writing, waitressing, stenography—and though I succeeded pretty well at them all, somehow I wasn't happy because all along I knew that it was a cowgirl that I was cut out to be, and I was meant to ride the range."

That is a quotation from a letter written in hope and good faith by an American girl. On February 18, 1938, Audrey Jane sat down in her Chicago home and addressed her petition to Harvey Witwer, then co-manager of a dude ranch at Greeley, Colorado. It has lain in the Witwer family archives, with a number of similar specimens, ever since.

It won't do to laugh at Audrey Jane. There are too many like her. Every year thousands of starry-eyed tenderfeet bombard the American ranchman with letters, telephone calls, and personal visits. They want a job and a vacation combined, in the saddle and on the range.

"You'd be surprised how many letters we get," says Earl Monahan of western Nebraska, "and how many of them are from girls. They all want to come out and ride. We generally throw those in the wastebasket."

Wallis Huidekoper, who was a tenderfoot himself when he came to Montana in 1886, is a little gentler about it: "By every mail we hear from somebody who wants to send his son out to be a cowboy. We finally had to make up a stereographed letter explaining that we had no openings at the moment."

The really amazing part of it is the fact that these deluded dreamers sometimes—just sometimes—make their dreams come true. Peter Stites, a red-headed, six-foot boy from Cincinnati in his senior year at Williams, has just finished his second summer on Clyde Buffington's place outside of Gunnison, Colorado, as this is being written. He saw Clyde's name in the *Livestock Journal* and wrote to ask for work. "Give me the hardest job on the ranch," he begged. "I want to see if I can take it." To Clyde's surprise he took it, and came back for more.

In the foothills of the Rockies back of Golden, Colorado, Joe McNamara thinks he is having the time of his life fixing fence and making hay for George Green. He has no cattle of his own, but he lives on his own place with his wife and baby and makes out by helping the neighbors. Born and bred in the East, Joe first saw and loved the West when he came out as a soldier. Before the war he made his living as an acrobatic dancer. "There are lots of such cases," says George Green.

Another one for Believe-It-Or-Not is the case of Dimitri Poutiatine who is now up in the mountains south of Monte Vista, Colorado, learning what to do with a posthole digger and a manure fork. Russian born, Dimitri was going to school in Belgium when the German occupation caught him. He escaped to France and came to America with his parents in 1941. During the war he served with the armed forces in Japan as an interpreter. Afterward he came back and studied International Law at Harvard. When he was nearly ready to graduate, he made up his mind that he wanted to learn about—of all things—ranching. He knew as much about cows as a puncher does about Plotinus, but he wrote to the secretary of the Commercial Club at Monte Vista asking what he could do. The secretary referred him to Luke McOllough, a substantial local cattle raiser who loves his fellow man. Luke let him have it with both barrels—told him about the jobs he would have to do, digging, hauling, irrigating. "I'll give you $125 a month and a house if you want to try it," he promised him, not really expecting the matter to go any further.

Dimitri telegraphed right back: "I accept your offer with pleasure."

In April he arrived at Monte Vista with his wife, his Master's degree from Harvard, his European background, and his unusual name—all complete. After six months the McOlloughs are delighted with him.

"We like him very much," writes Mrs. McOllough. "Also

Irene, his wife. They lived here with us for six weeks, until they could get their house ready in the hills, about fifteen miles from our ranch, which was a job, but they neither are afraid of work. . .

"Lucas really likes him very much as he really wants to do everything just the way he's told, and doesn't have to be told but once how. He has done a little bit of everything such as driving tractors, irrigating, working in the hayfield, on the spud sorter and even to digging postholes. Tomorrow he will help bring the cattle from the hills. They will have to ride two or three days to round them up."

Why do people do these things? What strange bug bites these otherwise normal Americans and drives them out to sweat and struggle at tasks for which they are totally unprepared? Is it because of a subconscious nostalgia for our vanished frontiers working in us like some spiritual yeast?

There is no simple answer. But of one thing we can be sure—with the assistance of Hollywood, the radio, the pulp magazines, and songs about tumbling tumbleweeds and riders in the sky, we have made a myth out of life on the range.

It is not a bad myth in its way—harmless, exciting, and possibly beneficial to our crumbling moral standards, for it never allows the wicked to win. The chief objection is that it wipes out fifty years and more of history as if they had never been. It suggests that somewhere beyond the purple hills life goes on as it did in the eighties; that a heroic and independent breed still follows the Code of the Range; that tall cavaliers still ride, rope, sing, and perforate their wicked enemies as skillfully and nonchalantly as of yore.

The stories and songs never even hint that the old cowhand from the Río Grande does most of his business in a pickup truck these days and spends the best part of the summer in the hayfield, though such are the facts.

The cattlemen themselves do their full share in keeping the myth alive, even while they heap scorn and insult on writers, dudes, and tourists who come looking for relics of the Old West. In 1948, Dan Casement, a Princeton graduate who ranks as a sort of dean among cattle ranchers and is a very thoughtful and articulate man, stood up before a meeting of the American National Livestock Association at Phoenix, Arizona, and delivered himself as follows:

You do not represent a business system or a political organization. You are a social class, typifying a way of life, a fraternity of ideals, that preserve the best in American lore, that unify in a single code of citizenship the traditions of our forefathers for freedom, independence, opportunity, resourcefulness, and rugged individuality.[1]

"The West has tended to see itself as it has been seen by others," adds Carey McWilliams, "and what the others have seen, of course, is the myth."[2]

The myth explains a good many things. It explains, for instance, the proud bearing (some people call it arrogance) of a booted cattleman in the presence of men with shoes on. He knows he belongs to a tough and independent breed, and he knows that they know it. He senses that they are appropriating a small share of his pride in his heritage; that his fellow Americans—mechanized, regimented, obsessed by security—see him as a symbol of their own beginnings, of the unfenced life of the Frontier of other days.

And how all of us love to play at being the thing we dream of! How we dress up for Frontier Days celebrations and enjoy riding the stagecoach in the parade! How we pour money into the horses and equipment necessary for membership in the Sheriff's Posse! How we sweat as we elevate our pursy persons into the saddle and groan as we unload them at the end of the ride! It takes great devotion to make a man gall and exhaust himself in this fashion.

Middle-aged burghers in the Sheriff's Posse are not the only ones affected. Next time you go to a rodeo, look at the shapely young things in levis, checked shirts, and feminized Western hats (at the last rodeo I saw they were going in for little curly-brimmed green straws) and ask yourself what they are saying, not with their lips, but with themselves.

[1] Reported by Edward N. Wentworth in an address before the Society of Westerners, and printed in *The Cattleman* for March, 1948.

[2] "Introduction" to *Rocky Mountain Cities,* edited by Ray B. West, Jr. (New York, W. W. Norton, 1949).

It goes something like this: "I am a sophisticated young woman, as you very well know, and I shall be dazzling in my new formal at the country club dance tomorrow night. But for the moment I choose to show you another, simpler self. This is maybe the real me—a healthy, companionable, unpretentious person who is much closer kin than you ever suspected to my Grandmother Purity Peebles who came to these parts in a covered wagon."

Thus on the one hand we have the myth of the cattle range, with its roots in historic reality and its top in the clouds. But beside it we have to place a darker picture—and that is where we begin to realize that the West has two faces. For the last few years have etched on the national mind a picture of the American cattleman which places him somewhere between a sneak thief and a robber baron. Partly because of our concern over the plundering of our planet, partly because of efficient propaganda from various agencies with which the rancher has to deal, partly because of ill-advised moves on the part of prominent stockmen and their representatives, the country is full of honest citizens who damn cattlemen collectively as selfish opportunists who rob their own ranges, despoil the public domain, and grow callously rich at the expense of the rest of us. Bernard De Voto fired the opening gun:

The cattlemen came from Elsewhere into the empty West. They were always arrogant and always deluded. They thought themselves free men, the freest men who ever lived, but even more than other Westerners they were peons of their Eastern Bankers and of the railroads which the bankers established to control their business. With the self-deception that runs like a leitmotif through Western business, they wholeheartedly supported their masters against the West and today support the East against the West. They thought of themselves as Westerners and they did live in the West, but they were the enemies of everyone else who lived there. . . . Nothing in history suggests that the cattlemen and sheepmen are capable of regulating themselves even for their own benefit, still less the public's. . . . Cattlemen and sheepmen, I repeat, want to shovel most of the West into its rivers.

This blast appeared in *Harper's Magazine* for January, 1947. On July 26 and August 9 of that year, *Collier's* moved in from the rear with two articles by Lester Velie called "They Kicked Us off Our Land." Again the cattleman emerged a bloody mess.

The stream of criticism let loose by Messrs. De Voto and Velie immediately swelled into a flood. Sportsmen's organizations, conservation agencies, even ladies interested in preserving our wild flowers, picked up the stiletto or the maul and attacked the cattleman furiously. Puzzled and infuriated, the cattleman fought back as best he could—through lobbies and pressure groups in Washington (thereby bringing on himself more damnation) and, less effectively, through the press. The battle still goes on, and it is not just a war of words. Men on both sides are angry enough to fight over grazing matters at this moment. Many a ranchman's livelihood is at stake. And posterity is shaking its spectral head over the exploiters who are said to be reducing our American land to a sterile desert.

To make matters worse, the cattleman is temporarily prosperous. After eighty years of going from riches to rags, he has reversed the process and gone from rags to riches. To paraphrase Churchill, Never have so many made so much out of so few. For twenty years he lived in what is sometimes called the Buffalo Chip Period when he fought the Indians, pushed into new country, and used what delicate-minded cattlewomen still refer to as "prairie fuel." Then he moved into the Cow Chip Era, which lasted for another twenty years as the ranch business got on its feet. During this time his campfires were nourished by prairie fuel provided by the cow herself. Now we have had ten or twelve years in the Blue Chip Epoch. The rain has fallen over much of the range. Prices have skidded upward to what seem like astronomical heights to the men who sold their half-dead critters to the government in 1934 for twelve dollars a head. The cattleman is out of debt and has bought a new Buick.

This prosperity has not made him popular with his fellow Americans. The man who has a dollar—we may as well admit it—never looks very good to a fellow citizen who has only fifty cents. For this reason Mary Kidder Rak, who lives in Rucker Canyon thirty-five miles out of Douglas, Arizona, and has written two good books about cattle people, says that the cattle-man's case is hopeless. His only chance is to sit quietly and at-tract as little attention as possible, because nobody is going to love him anyhow. From the first, stockmen have had to control a lot of land in order to make a living at all. Even a small operator in the arid regions of our country has always had to be master of an impressive number of acres, even though those acres would grow nothing but cactus, rattlesnakes, and a few skinny cows. He has fifty thousand acres. He must be wealthy. And since wealth inevitably corrupts, he is bound to be a crook.

There are the cattleman's two faces. To many he is the per-sonification of the American ideal, the repository of all our an-cestral virtues. To others he is a menace. Our house is divided against itself on this issue, and it is important to find out which side is more nearly right. Our dreams and ideals, as well as our national economy, are involved. And how can we fail to be con-cerned when our old friend the cowboy is in such serious diffi-culties? If Mr. De Voto and Mr. Velie are right, we have a lot of damage to repair. If they are wrong, we have a lot of apologizing to do.

It occurred to the members of the Rockefeller Committee at the University of Oklahoma in the fall of 1948 that under the circumstances a report on the American cattleman might be in order, provided that some guileless character could be found to undertake to make it. He would have to risk a trip into the haunts of the predatory stock raisers and hope to get back with his notes. He would need to make friends with them, eat with them (if asked), sleep in their houses, and watch them at work. When he got through, he might be able to give a reasonably de-tailed and accurate account of their way of life, education, ideals, and ambitions.

I was the guileless character who accepted the assignment to invade the American cattle country to see what there was to see. From February 2 to June 23, 1949, I drifted through the range lands from the Río Grande to the Yellowstone, from the Nebraska Sandhills to Great Salt Lake. I traveled from end to end of what was once the Cattle Kingdom and is still the heart of the cattle country, learning everything I could.

In these pages are set down much that I found out about the people who raise beef for a living—not about the trail driver and the pioneer, but about Archie Sanford and Al Favour and Alfred Collins and Alvino Canales and a hundred more who still saddle up early and go to bed late on the American cattle range. The object is to look at all of them—old-timers and newcomers, big ones and little ones, good and bad—and see what makes them tick.

C. L. SONNICHSEN

El Paso, Texas
  July 5, 1950

Cowboys and Cattle Kings

LIFE ON THE RANGE TODAY

PART I In Blood and Bone

T

1. Gone with the Longhorn

HERE IS A STORY, which may or may not be true, about old Captain Jones who use to own about half of South Texas. He accumulated many millions of dollars and a variety of real-estate holdings, including the Nueces Hotel in Corpus Christi.

Shortly after the hotel was built, a lady of excessive refinement registered at the desk and turned for a look at the impressive lobby. She saw an old man sitting in a rocking chair, at peace with himself and the world, vigorously chewing tobacco. Every now and then he would release a high-pressure jet of tobacco juice into the surrounding atmosphere, letting the drops fall where they would. Immediately a flunky would rush up with a mop and take care of the situation.

The lady was scandalized. She said to the clerk, "That is the nastiest man I ever saw in my life. He is actually spitting on the floor. Why don't you put him out?"

"Madam," replied the clerk, "I work for that man. He owns this hotel, and he can spit on his own floor any time he wants to. It will take a lot of doing to put him out."

You hear such tales about the old-time cowman. Always colorful, always frankly and sometimes embarrassingly himself, he hung on to his old habits whether he was broke or up to his hocks in money. The cattle country still likes to spin yarns about him—how Colonel Goodnight, who was a practicing Baptist, forbade his cowboys to play mumblety-peg on his ranch; how Shanghai Pierce put up a statue of himself in the middle of his biggest pasture; how Theodore Roosevelt got pitched off a horse and said to the cowboy who offered to take over, "No thanks. I know you can ride him. I've seen you ride. What I want to know is whether *I* can ride him."

5

These things happened in what M. M. Kelso calls "the Robin Hood days of the business,"[1] when civilization had not yet knocked the corners off the cattleman—when he was still untouched by feminizing refinements such as Cadillacs, easy chairs, bathrooms, and company towels. Good or bad, he was an honest-to-God person who stood on his own feet, thought his own thoughts, and lived his own life.

His backgrounds were apt to be as varied as his personal peculiarities. This was particularly true in the northern part of the range country. In Texas the stock was prevailingly Southern and the men were usually brought up in more or less close contact with the cattle business. In the north they came from everywhere and had worked at everything.

The cattlemen were a mixed lot. Some were disappointed miners from Last Chance and Virginia City, like Joe Scott and Granville Stuart; some had come west to fight Indians, some for their health and some to hunt the buffalo. The sporting type was represented by Theodore Roosevelt and the sparkling Marquis de Mores, both of whom ran cattle in western Dakota in the district controlled by the Montana Stockgrowers association. There were only a few "cattlemen" who had had experience in raising cattle before. J. W. Toohey started in the business when the vaudeville company with which he was associated disbanded at Miles City. In a group of 56 biographies of cattlemen living in Chouteau county in 1884, it is revealed that 18 had been miners, 10 general farmers, 8 traders and freighters, 5 railroad hands, 3 soldiers, 3 clerks, 3 hunters and trappers, 3 steamboat hands, 2 grocers, 2 lumberjacks, 2 butchers, 2 iron moulders, and 2 saloon keepers. One each at some time or another engaged as a sheepherder, factory hand, interpreter, tinsmith, wiremaker, blacksmith, miller, lawyer, mounted policeman, Indian scout, school teacher, photographer or vine dresser in France. Only four had previously been engaged in the cattle business.[2]

What could the individuals in such an assortment as this have in common? Not much, probably, except the qualities

[1] "Present Day and Prospective Ranching," June 22, 1948, MS. Mr. Kelso is head of the Department of Agricultural Economics at Montana State College.
[2] Robert S. Fletcher, *Organization of the Range Cattle Business in Eastern Montana, Bulletin No. 265*, Montana State College Agricultural Experiment Station, Bozeman, Montana (June, 1932), 3.

needed for survival on the frontier: courage, resourcefulness, endurance—the ability to take it, and to "tough it out." A sense of humor helped, too, and a willingness to go ahead and do what had to be done, whether there were any rules for it or not.

The books and articles which reveal the essential nature of the cattleman as he used to be number into the thousands, but the old man himself has almost disappeared. Only a handful of Model T stockmen are riding rocking chairs today on the front porches of the town houses they have retired to. The things they stood for, good and bad, are still present in the subconscious mental operations of today's cattle raisers, but as memories rather than realities—like the vanishing trail of smoke across the prairie after the train has disappeared. It could not well be otherwise, for revolutionary changes have come into the beef-raising business since the first herds dipped their muzzles in the Yellowstone.

First came barbed wire to put an end to the open range, and with it the carelessness and piracy of the days of free grass. It finished the old-time cowboy and his methods of handling stock —the cattle drive, whose chronicles are the Odyssey of the Cattle Kingdom—the long-legged, thin-flanked, fearfully weaponed longhorn, full of brimstone and vinegar.

Hard on the heels of the barbed-wire salesman came the hoe man with a plow tied to the tail of his wagon. Far out on the arid plains, where no farmer should ever have showed his face, the grama and buffalo grass went under in the furrow.

The invasion of the farmer was backed up by the development of irrigation. On the parched plains of West Texas the cattleman cannot produce much supplementary feed, but in the mountain valleys of Colorado, in the river bottoms on the plains, in the areas where shallow water can be sucked up by powerful pumps, every scrap of irrigable land is producing—a change which tends more and more to make farmers out of cattlemen.

Along with irrigation came mechanization. The windmill and the mowing machine came early. More recently good roads and motor vehicles have caused the ranchman to depend more on the pickup than on the horse. And even when horse work has to be done, a horse trailer takes man and mount to the place of operations. The cow ponies themselves have undergone some changes. J. Y. Crum says his horses enjoy riding in the trailer and jump in without being urged. Even in the ranchman's home the old ways are no more. In 1947, Texas reported that rural electrification had reached 141,816 rural families (farmers and ranchers), that the number was increasing at the rate of 3,000 a month, and that 4,807 bathrooms had been installed in rural homes during that year.[3] Now electric lights replace the kerosene lamp; radios, sewing machines, and separators are run by electric power; and deep-freeze units put fresh meats and vegetables on the table at every meal.

With these gadgets have come other improvements in ranch methods—better bookkeeping, pasture clearance, long-term planning, soil conservation, and expert help from the United States Department of Agriculture. The ranchman is a thousand times better educated for his job than he used to be. Even his social and cultural life has been transformed by easy access to towns and cities and to the homes of his neighbors.

The cows themselves have changed along with everything else. Improvement of livestock has done as much as anything to make the ranch a different sort of place from what it was fifty years ago. Fine Herefords, Aberdeen Angus, and Shorthorns have replaced the longhorn and developed a new breed of men to work them. The days when John F. Yearwood could be suspected of insanity when he bought a six-hundred-dollar bull are gone forever, and every breeder nowadays knows that he makes more money by breeding better stock. Where lean and leggy critters once paced the pastures and broke for the brush, bovine behemoths now ruminate in massive calm.

And the way those expensive animals are soothed and pampered would make a fairy-tale princess envious. "What happens," asks the 1948 *Annual Report* of Texas A. and M. College, "when good crossbred cattle are put on improved, fertilized pasture, given mineral supplement, sprayed with DDT for flies, treated with rotenone, phenothiazine, BHC and hexachloroethene, and rotated from pasture to pasture with the seasons? A good many

---

[3] *The Seventh Year of Record Production, Annual Report* of Texas A. and M. College Extension Service, 1947, pp. 4, 13.

Texas stockmen already know: Close to 100 per cent calf crop and calves up to 200 pounds heavier at weaning time!"

In other words, modern methods make money, and with more money the cattleman has moved out of the era when he expected to be wiped out every few years, never got out of debt, and engaged in a lifelong struggle with cattle diseases, bankers, the elements, panics, distance, and discouragement. According to the report just quoted, the average cash income of farm and ranch families in Texas in 1930 was $1,069. In 1947 it was $4,-725. "It seems likely," the statisticians conclude, "that historians will define this era of change as the Agricultural Revolution."

So the old cattleman has gone—shoved out by barbed wire and windmills, by account books and pickup trucks, by DDT and hexachloroethene. But we lost a good man when he went away. He deserves a salute from all who love independence more than security, courage more than conformity, endurance more than the gift of grab. But does he get it? No!

Nobody talks now about the real service these old-time cattlemen did in bringing the most inhospitable part of the continent into profitable use. What we hear about is their exploitation of free land, their misuse of the range, their forcible discouraging of plowmen, their arbitrary methods of insuring their own safety and profit. These unpleasant habits, which always seem to go with the settlement of a new country, are being dug up and thrown in the faces of their descendants and successors. "This is the way cattlemen have always been," say the critics, with a heavy sneer.

To Dee Linford they have always been savages with no redeeming feature. "On the side of the homesteader was moral and legal right," he says. "On the side of the stockman was MIGHT—and absolute control of the territorial government." When the cattlemen organized, they did so, he thinks, for dark, conspiratorial reasons:

Within Wyoming itself, this powerhouse clique has dominated every legislature and dictated policy in varying degrees to every governor but one since its inception in 1873. . . . In 1884 the National Cattle Growers Association was established in Chicago, on the Wyoming model, with Wyoming stockmen at its head. This "association of associations" was later to become the American National Livestock Association, the super cowman lobby which still seeks to accomplish on a national scale the things which the Wyoming association has put over locally in Wyoming.[4]

Let Mr. Linford forget his partisan economics and his bad-tempered liberalism for just a minute and think what it must have been like back there. We can see them yet, those vanished old-timers, pushing their herds along the cattle trails, eating the dust of the drags, shivering in the brutal cold of the northern winters, fighting the wild rivers, standing off the unpredictable Indians and the all-too-predictable settlers, living on tough beans and tougher meat, lonely and tired and lousy.

Oh, yes, it was a romantic life—to read about. And even if a man had been through it all, the delicate, many-colored fungus of the years eventually covered the hard realities and he began to think that it was a pretty good time, too—particularly if he read a few books and attended the meetings of the Old Time Trail Drivers Association.

But in his heart he knew all the time that it had been a tough business full of trouble and sudden death, of ruthless ambition and great hardships, of the long frustration of men without women, without home, without any certainty in the future. He admitted to himself that many of his old companions were rough and ignorant frontiersmen who knew cows and not much else, whose biggest and sometimes only virtue was brute courage—a profane, violent, hard-drinking, pugnacious crew who appropriated other people's cattle without compunction and hanged anybody who stole theirs.

Yes, he admitted it. But he refused to take the final step and lump the old-time cattleman with the other barbarians of history. He knew that along with the hell-roaring old reprobates there were deeply religious men. Along with the sticky-fingered rascals who couldn't have been trusted with a busted cinch, there were rigidly honest citizens who wouldn't have touched a stolen

---

[4] "Cheyenne," in *Rocky Mountain Cities*, edited by Ray B. West, 119, 120, 124.

steer with a ten-foot pole. To him it was a simple fact that those vanished cattlemen were like other men—some bad, some good, but mostly well-meaning citizens who sometimes made mistakes. They were the children of their times, and could hardly have been otherwise than they were.

He wished somebody could make it plain to the bright and bitter young men of the new time that, in spite of what was wrong with them, they had a steadfastness and courage for which this generation had better show some respect.

T 2. New Model

HE SQUARE at Columbus, Texas, has echoed to the tread of boot heels for almost a hundred years. Under the immense live oaks Shanghai Pierce once traded cows with other Gulf Coast cattlemen. Colonel Bob Stafford's magnificent old house takes up half a block on the south side. Around the big yellow courthouse the gunfire of two cattlemen's feuds has shattered the peaceful atmosphere. Seventy years ago the wide prairies swept south and east to the Gulf without a fence—with hardly a tree—and the cattle kings of yesterday ruled their empires in the old high-handed way.

It is quite different now. German and Czech farmers outnumber the cattle growers on the bank corner of a Saturday afternoon. Trucks loaded with oil-field machinery clink and clatter past. Rice farmers, gravel-pit operators, chicken ranchers, and truck gardeners fill the barbershops and hunch over the counters, elbow to elbow, in the cafés. Columbus is no longer a real cattle town. And yet there are probably more cattlemen in it today than there ever were before in all its history. In fact, it seems that every other citizen is dabbling or plunging in livestock.

Start with W. H. Mikoe and Ellis Miller, presidents of the two banks. They have cattle. Dr. H. C. Miller, one of the dentists, has cattle on two places. Frank Talbert, who runs the Chevrolet Motor Company, has a herd. Marley Giddens, superintendent of schools, has forty-four head on leased land. Tanner Walker,

11

who manages the Live Oak Hotel, runs a hundred purebreds and another hundred head of grade stuff. He keeps a Negro hand on the place and doesn't spend much time there himself, except in the busy season; but he thinks of himself as a cattleman first and as an innkeeper second for the simple reason that he makes more money out of the livestock than out of the Live Oak.

More or less the same thing is happening everywhere. Gone are the handlebar mustaches and the men who wore them, and a mixed breed has stepped into their boots. It is a question in some people's minds whether there are any real cattlemen above ground. Breeders, feeders, and stock farmers—yes. But few if any of the old stock.

"You may think real cattlemen or cowboys still exist," writes Mrs. C. T. Traylor from Cuero, Texas. "I assure you they do not."

"There are not so many real cattlemen," adds Mrs. O. L. Shipman, historian of the Big Bend country. "They are ranch people and not cattle people. There will never be a ranchman with the romantic background of the cattleman."

True enough, in many of the old haunts of the *vaquero* and the longhorn the cattleman is practically extinct. In the Fort Stockton region the years of drought between 1944 and 1949 left many ranchmen no recourse but to feed sheep for a living. Del Río, once a cattleman's stronghold, now advertises itself as the Mohair Capital of the World. Farther east on the coastal plain, cattlemen like Lester Bunge of Altair are putting part of their land into rice—when they can spare a little time from their oil wells.

Not for a minute, however, do these diluted cattlemen give up the notion that they belong to the old breed. Step into the bank at Luling for a talk with Miller Ainsworth and see for yourself. Ainsworth is not merely an oil millionaire, banker, and trustee of the Luling Institute. He is also a general in the National Guard and spends half his time organizing military activities in his district. At the same time he owns a 10,000-acre ranch where his grandfather settled on a Spanish grant in 1836. Ask General Ainsworth if he is a cattleman, and he will reply emphatically that he is.

12

New Model

At Albany, Texas, County Agent W. C. Vines has leased five hundred acres northeast of town and has the place stocked with sixty head of good commercial cows. He loves cattle and is ambitious to have a fine herd of his own some day. In the meantime he covers the county in his pickup, giving advice when it is asked for and helping with the ranch work when his friends are short handed. Mr. Vines is too modest a man to go around bellowing that he is a cattleman. But why shouldn't he?

There is no need to multiply examples. All over the range country hundreds and thousands of businessmen, greenhorns, and dudes are in the game. They are cattlemen, too, after their own fashion, though their time may not be long. Bunkhouse sages and fence-rail philosophers regard them as merely temporary.

"There are lots of boys in the cattle business now who don't know all the angles," comments Eddy Phillips, secretary of the Montana Stockgrowers Association. "Those who came in since the drought of 1936 have seen nothing but a long series of rises in the price of beef. The years have all been good years. Any fool could make money. When the bad times come, some of them are going to get hurt. The legitimate rancher keeps his house in order and will get by all right. He owes less money than ever before, hasn't gone in for any wild land buying, and knows how to get ready for trouble. But I'm afraid for some of the others—especially the rich boys."

Goob Saunders, cattle inspector at Dickinson, North Dakota, thinks these new ranchers will be victims of overconfidence if they don't watch out. "All the young fellows who have come into the business in the last five years have made lots of money —sometimes more than their fathers made in their whole lives. So they think there was something wrong with their fathers. If they stay in the business, they will go broke, too. And when the hard times come around again, they'll know what it's all about. Government subsidies won't help. No matter how much you support prices, what can you do when it don't rain?"

If greenhorns and amateurs have cracked the solid front of the cattleman, the farmer has made a breach that you could drive a cattle truck through. It is good business and the best

13

kind of insurance for a ranchman to raise his own feed, but the practice goes further than that. Mixed farming and ranching layouts are extremely common. In fact, it has become so difficult to draw the line between farm and ranch that only the most courageous agricultural economists will attempt it. The staff at New Mexico A. and M. College distinguish between the stock ranch and the stock farm. A ranch is not a ranch, they say, unless it has "ten acres or more of grazing land to each acre in crops."[1]

The cattlemen themselves have long got over feeling any shame because of their merger with the hoe man, though sometimes a remnant of the old prejudice will crop out. Huling Means, Jr., for instance, has begun to transform his ranch in the Estancia Valley of New Mexico into a mixed farm-ranch enterprise, thanks to a fine flow of subsurface water coming up through his new pumps. The Meanses are cattle people from away back, and Huling's mother takes on half-humorously over what is happening to her son. Ask her what objection she has to seeing him get rich off his irrigated acres and all she can say is: "But he's a cattleman!"

Almost as startling as the transformation of the cowman into a farmer is his amalgamation with the sheepman. Grandpa would never believe his eyes if he could come back and see what has happened. In some sections the country has gone over almost entirely to sheep raising—in southwestern Texas, for instance, and in northwestern Colorado.

Norman Winder, of Craig, Colorado, a personable and gentlemanly sheepman who has been president of the American Wool Growers Association, tells how it happened in his corner of the state. "The bad times of 1919–20 crippled a good many cattlemen. That was a terrible winter. Many cattle were lost and many men went out of business. The depression got some more of them, and this is sheep country now. However, most sheepmen raise a few cattle. The cattlemen don't usually raise sheep, though many of them ought to. Service brush and buck brush are good sheep browse, but not good for cattle. The cattlemen

[1] Byron Hunter, P. W. Cockerill, and H. B. Pingrey, *Type of Farming and Ranching Areas in New Mexico*, Part I (State College, New Mexico, Agricultural Experiment Station of the New Mexico College of Agriculture and Mechanic Arts, May, 1939), 7.

OLD-TIME COWBOY
"Scandalous John" Selmon listening to a tall one

Photograph by Charles English

don't put sheep on their brush pastures, but it isn't because of any prejudice. They just don't know how to raise sheep."

A man who has considered the sheep situation carefully is Guy S. Combs, Jr., who ranches in the broken country south of Marathon, Texas, at the top of the Big Bend. Guy is six feet three, with a pair of lively blue eyes and a nose like the jib sail on a boat. He also has an inquiring mind and a degree in geology from Stanford University. He has turned his share of Grandfather Combs's pastures into a sheep ranch—and he knows why.

"The property which came to our side of the family originally had ninety sections after we sold off the poorer parts," he explains. "It carried 2,500 head of cattle. As the country went down, the weight of the calves dropped, showing overgrazing. Carrying capacity fell to 2,000–1,800–1,500 head. During the drought it dropped to 900 head. That was about fifteen years ago. It has good grass cover now, but the grama grass has been killed out.

"A lot of the country washed away, too. The old wagon road to the river ran down Maravillas Creek bed through Maravillas Gap right out in front of our house. The wagon ruts, as usual, were invitations to erosion. Cattle and sheep made side trails up the hills, and those washed also. The range went down—and then came the brush. The whole place is grown up now so that it would carry only a handful of cattle. The only thing to do was to bring in sheep and goats to eat what grass there was and browse on the brush.

"But there again we ran into trouble. The sheep range farther east—around Rock Springs and Sonora—began playing out, and about 1935 sheepmen from that country crowded in here, offering more money for leases than cattlemen could afford to pay, as much as thirty-eight to forty cents an acre. They put up a plausible story about how the sheep would actually improve the range. They would eat grass, seeds and all, through the winter. Their droppings would fertilize and spread the seed. This would have worked with a reasonable number of sheep. But the sheepmen had other ideas about that. Of course the contract for the lease reads so many sheep per acre, in black and white, and signed by both parties. But try to hold a man to it! If you go in and try to count, you make trouble. And if you decide to do it, you'd better be sure. Consequently a lot of sheepmen never paid much attention to the contract. On one ranch of thirty-three sections a visiting sheepman ran from twelve to eighteen thousand sheep and goats. The place is carrying 2,200 breeding ewes and 300 goats now.

"Ever since the first rancher came into the country the range has been continuously pastured and the grass has never had a chance to reseed, so we have a tremendous job of rebuilding to do. Contrary to the usual practice I take my stock off the mountains in the summer and put them on my worst land. The lambs are lighter, but the pastures have a chance to rest and reseed, and in the long run I think my way will turn out to be right. I've been on the ranch two and a half years and made money every year."

Men like Guy Combs have helped to heal the breach between the cowman and the sheepman. The need for making common cause in pressing legislation, fighting the Forest Service, influencing taxation, and so on, has almost finished the job. Combination ranches running both cattle and sheep are common everywhere. In fact, sheep have proved so valuable that they are sometimes called "mortgage raisers," as are hogs in the corn belt. Only among the older cattlemen does the ancient prejudice against the "stinking sheep" and his owner still appear, but when it does, it can be bitter.

"Some of my best friends are sheepmen," says one old-timer, "but ordinarily I don't care much for them. I've run some sheep myself, but it was in self-defense. The difference between a sheepman and a cattleman is that the cattleman lives on his place and takes care of it. The sheepman lives in town because he is ashamed to be seen at his sheep headquarters—usually a heap of scrap iron, wire, and junk. A disgrace! They are all dumb and ignorant. The Lord makes the grass grow. Sheep can't help producing wool. It doesn't take any brains to collect the money.

"Even the herders are a mess. I wouldn't sell a horse to a sheepman. They put a saddle on it—an old hull, with a knotty old blanket underneath—don't take it off for three weeks, and when they do, the hide comes with it. Old, worn-out cow horses are ideal for sheepherding, where all a man does is sit up there all day. But the horse dies of a broken heart. I would rather shoot one than sell it to a sheepherder. I pension my old ones off and let them enjoy a little peace and free grass.

"Sheepherders even abuse their dogs. They cut the band down—don't need so many dogs—beat and throw rocks at the ones they don't want—run them off. The result is bands of homeless dogs."

Thanks to the establishment of the Grazing Service, many of the causes of friction between sheepman and cattleman have been removed, and now each group seems inclined to speak of the other with tolerance, if not with deep respect.

"Before the passing of the Taylor Bill," summarizes George Snodgrass, "there was no law to prevent anyone from turning loose a thousand or a hundred thousand head of cattle or sheep on the range, and the result was that it was overdone. The cattlemen as a group had always tried to get some form of lease law passed to control the grazing on the range and the sheepmen as a group always opposed such a law, for they already had the advantage by being able to starve out the cowman. A sheep can graze so much closer than a cow that the cow can't compete, and will starve. Since the passage of the Taylor Act, the feud between the cattlemen and sheepmen has disappeared. Sheep numbers in Wyoming have been reduced by about 50 per cent and the range is starting to improve. So is the relationship between cattlemen and sheepmen as they work together on mutual problems."

Leaving aside the cattle raisers who have deserted to the sheepmen or been seduced by the farmers, we can turn to the ones who make their living entirely from cattle and find that there are cattlemen and cattlemen—and cattlemen. For the business is much more involved than when Grandpa branded mavericks and double-wintered his herd on Dakota grass. Like so many vocations, beef production has become a complex of interlocking specialties.

There are really four steps in the meat business, says a recent pamphlet,[2] and consequently there are four kinds of cattlemen.

[2] American National Livestock Association, *Meat on the Nation's Table* (Sheridan, Wyoming, n. d.), 2–3.

Number one is the pure bred man who comes first as "the foundation of beef-animal production." Then we have the commercial breeder "who produces the calves which are the offspring of commercial herds of females and purebred sires." Third is the pasture cattleman (the one we usually call a "rancher") who makes meat by feeding grass and hay to calves and yearlings. And finally comes the feeder or finisher who gives the animal the ultimate polish before slaughter.

Many real ranchers are in the purebred business, though most purebred men throughout the nation might better be called stock farmers. Either way there is a special kind of excitement about them and their business. They have the glamour, the money, and the showmanship. They play for high stakes, go up like skyrockets, and come down like meteors. Yet on them as a foundation the whole cattle industry rests. For fifty-thousand-dollar bulls and blue-ribboned champions are just means to an end—the end being the siphoning of better blood into the cattle on the range. Better blood means more pounds per animal, quicker development, and a lower percentage of loss in dressing-out. Place a picture of the jack-rabbity longhorn of sixty years ago alongside one of a low-slung, globular modern beef steer, and the difference is as obvious as a Hereford's rump.

This is the gilt edge of the cattle business, involving investments of hundreds of thousands of dollars, sumptuous ranch houses, expensive show herds, and million-dollar dispersal sales. It is the side which brings Big Business to the range, and it is not for amateurs to play around with. There are successful men who started small—maybe with a 4H-Club calf or two. But the day when that was possible is rapidly passing. The ones who go into the purebred business now are likely to have inherited the place, married a rich woman, or struck oil. Clyde Buffington estimates that $75,000 is about the minimum capital necessary for breaking into the business.

But money is not enough. A purebred man has to have hair-trigger cow sense, a gambler's courage, and all sorts of know-how. He has to have his blood lines at the tip of his tongue, develop a flair for publicity, and cultivate an ingratiating personality. His business is about as far removed from the ranching

18

tradition of the past as a pair of nylons is from buckskin leggings, and he himself is no more like the cowman of the eighties than a Chinese mandarin.

For one thing he has a completely new set of ideals. Where the old-time trail boss thought only of getting his steers to grass or to market, the purebred man waves his blue ribbons like a banner and shouts his war cry: "Let's improve the breed!" It is a sort of religion with him, and sometimes about the only religion he has. When he dies or retires, the final eulogy breathed over his departing form is, "He did much for American livestock."

Outsiders may wonder how much of this selfless ambition is real, and how much of it is a cover-up for crass capitalistic competition. Some of it may be a pious smoke screen, but the purebred men themselves do not talk about their efforts with their tongues in their cheeks. They are genuinely proud of disseminating the blood of the fine stock they have developed and feel that in this dissemination they have made a "contribution."

Richard Kleberg of the great King Ranch, for instance, is as proud as a peacock of his Santa Gertrudis cattle—"the only new breed ever produced on this continent to meet the needs of this continent." He says they put on two hundred pounds more than any other breed in four years and are therefore in demand in both North and South America. And that brings him to a purebred man's punch line. "Our contribution," he declares, "is hemispheric."

In the Cattle Kingdom, men like the Klebergs, Dan Thornton, Fred C. DeBerard, Bob Lazear, and Alfred Collins are the princes. The spotlight is always on them. Not merely do they have to be gracious and hospitable at home—they have to be known and respected from one end of the country to the other if they are to keep on in the lead. It is an all-or-nothing proposition. Consequently they are good spenders, great travelers, persistent advertisers, and mighty pleasant fellows. The sign at the gate quite often includes the words "Visitors Welcome," for every visitor is a potential advertiser, and they know from experience that the rough-looking man in the battered pickup may have a thousand dollars in his pocket to spend on a bull calf, if he can find one that takes his fancy.

Below the princely level, of course, there are the lesser peers, the squirearchy, and the commoners of the business—all the way down to the man who has one good bull and a couple of registered cows. It is hard to find a rancher these days who does not have a finger in the purebred industry somehow.

In this tricky and risky business commonplace personalities are a rarity. Some of the most colorful cattlemen in the country are raising purebreds. But they conform to no pattern.

Take Walter Scott of Chadron, Nebraska. He resembles the traditional cattleman about as much as he resembles Leopold Stokowski. He is a small, pale, earnest, high-strung man in his sixties who started out, an orphan, in the Platte River country with no particular chance to amount to anything. In his youth he heard about the "Table"—the big plateau in western Nebraska near the Dakota line, and came up to take a look at it. He never went back, and when he retired, he owned a section of that marvelous land along with a championship herd of Herefords.

His divergence from the common run of cattlemen appeared unmistakably when his health broke. He went to a specialist in Omaha and had an examination. The doctor tried to send him away without a final interview, but Mr. Scott stayed where he was until the examiner was ready to see him. Then he spoke as follows:

"Doctor, I don't know whether you have any Christian beliefs or not, but I just want to tell you this. Spiritually I am about as well prepared as I can be, and financially I can get ready in a very short time. How long do I have?"

"I'm glad you mentioned your spiritual preparation," the doctor told him. "I'm an elder in a church myself and I think prayer will do more for you now than I can do. I'll give you a year to get ready."

Mr. Scott went home and had a dispersal sale. Later he sold his home place because he did not want to see it run down in the hands of a tenant.

Then came the surgery. After the second operation he was found free of malignant symptoms. "Doctor," he asked, "do you think I might have ten years left?"

20

"Scotty," said the doctor, "you might have twenty."

Scotty felt good about that, but he was not too much surprised, for he is a very religious man and knew all along that miracles do happen. He was one of the ten ranchers and farmers who organized the Open Door Revival Association up there on the Table—a church with no denominational ties, which emphasized the old-time religion.

Mr. Scott went along with Mr. Gordon, the preacher, on many of his revivalistic tours to help with the music. "I sing a little," he mentions modestly.

Not many purebred men are known as gospel song leaders. Mr. Scott may well be the only one of his kind in existence. But the next man will be interesting in a different way. Dan Thornton of Gunnison is a handsome, dashing figure who wears his Stetson at a rakish angle, rates many columns of colorful publicity every year and has just served his first term in the Colorado Legislature. His neighbor Clyde Buffington, who lives farther up the valley of the Ohio Creek, is completely unobtrusive. With his thin face, alert blue eyes behind rimless glasses, and tolerant smile he looks a good deal like a kindly old professor; and in his deliberate, thoughtful talk, each word carefully chosen, there is a further suggestion of the scholar. He wears black laced shoes in the house and gets out in the cow lot in ordinary work clothes. Hollywood would never allow him on a movie set in the role of a cowman. Dr. C. R. Watson of Mitchell, Nebraska, would never get by the casting director, either. He is a big, fat, genial fellow, a successful surgeon, who is trying his best to retire to his seventy sections in the southern edge of the Sandhills. So far, he has not succeeded, but his election to the presidency of the Nebraska Cattlegrowers Association shows how he stands in his own state as a cattleman.

Some day a good book will be written, from the inside, about the queer quirks and peculiar practices of the purebred business. Buffington says he might write it. If he does, the work won't suffer from a shortage of vigorous personalities.

As essential to the complete efficiency of the cattle industry as the purebred man is the feeder, who finishes the grass-fed

21

calves and yearlings on pasture land or grain. The two techniques are entirely different. There are places in the country—particularly the Flint Hills extending from Pawhuska, Oklahoma, north through eastern Kansas—where the thin soil bottomed with limestone produces tremendous quantities of giant bluestem. John Joseph Mathews says that in good years you can ride through it and some stalks will touch your hands held straight out from your shoulders. To a stockman from West Texas, where a cow has to be a detective to find the next spear of grass, the Flint Hills are simply unbelievable. He can hardly imagine that there is that much grass in the world. But there it is, waiting, and every spring the trains roll in by the hundreds and unload carload after carload of young stock at sidings and small stations. The herds trickle off across the rolling, treeless hills or disappear into the brushy creek bottoms, and start piling on the meat at once. The Chapman-Barnard Ranch at Pawhuska unloads a train full of calves nearly every night during the shipping season. And so it goes, all over the Flint Hills area.

These feeders grow calves of their own, think of themselves as ranchers, and call their places ranches. They ride horses, wear chaps, and do not have the faintest intention of being classified as farmers.

On the plains of Kansas there is a feeder of another type who considers himself primarily a wheat farmer. But the tender green shoots of the winter wheat make wonderful pasturage for cattle and sheep, and thousands upon thousands are shipped in during good years to graze in the endless level stretches until spring is just around the corner. In February or March, after the cattle are taken out, the wheat puts its best shoots forward and makes a crop of grain in due time. This is old ranching country, where once the big outfits ran their cows by the tens of thousands. Now it is an almost terrifying monotony of vivid green wheat, with houses and barns sticking up occasionally like sore thumbs, and telephone poles marching along, incongruously erect in that universal flatness.

The grass and wheat feeders are far outnumbered by the feed-lot operators who drain off a large percentage of the calf crop from the range states every year. In Iowa, Illinois, Michigan, Missouri, Nebraska, and some of the Southern states, these men give the steers their final veneer of fat before they go to the stockyards. California is a great feeder state. So is Colorado. In eastern Montana the business is growing. Even in the Río Grande Valley near El Paso there are big feed lots.

A man who feeds as many head each year as anyone in the United States, according to his friends, is Warren Montfort of Greeley, Colorado. In thirty-odd pens on his compact place he has had as much as a million dollars' worth of cattle readying for market. He buys good stock from Oklahoma, New Mexico, Colorado, the Texas Panhandle—from anywhere and everywhere—and feeds them every mouthful they can hold.

"This is a factory," Warren Montfort declares. "We manufacture beef and nothing else."

The manufacturing goes on all day, all year, year after year, with feverish intensity. The men are throwing the feed at the animals all the time. Troughs are arranged in a line half a mile long on the outside of the pens, which makes the feed-throwing process simple. The trucks have a huge hopper behind the cab with a worm gear inside for mixing the ingredients of Montfort's special ration as the vehicle moves down the alleyways between the pens. A chute protrudes from one side and the mixture falls steadily into the troughs as the truck moves along. It is something to see that solid line of white faces poked through the fence at feeding time, all intent on their business of eating their heads off.

Montfort's mixture is his own private concoction but he admits using dehydrated alfalfa, barley, corn silage, corn, linseed meal, and soybean meal. What else there is in that powdery, pale green preparation is a mystery.

It is potent stuff, too, and every once in a while one of the steers bloats up, one side rising on him as if he had swallowed a lopsided balloon. When that happens one of the men gets out of the truck and chases him about the lot, whooping and waving his arms. If the animal can be forced to run around a little, he sometimes belches and gets rid of his gas pressure, like a baby being burped. If that doesn't work, more strenuous measures must be taken. Mostly the steers just go on gorging themselves

23

and covering their ribs inches deep in beef and tallow without missing a bite.

Nothing is wasted. If an animal does bloat up and die, he is sent off to a rendering plant. People come from away down toward Denver to buy the mountainous accumulations of manure—weedless and (if we may say so) pure. It costs them a dollar a load if they break their own backs—a dollar and a quarter if Montfort's boys do the loading. One year he sold $26,000 worth of fertilizer, which makes one wonder if in the long run Montfort doesn't make more money from cleaning his pens than he does from selling fat steers. "People go crazy over it," says Mr. Reynolds, Montfort's second in command.

The story of how this tremendous enterprise grew is one of those business epics that we used to call success stories. Montfort was a farm boy, not a rancher by birth, and he is not a rancher now, although the cattlemen could not do without him. About 1920 he and his father looked at the family checkbook, found it limp and anemic, and thought they would try feeding a few head of cattle. It was not a good time to start, and for a few years they had hard sledding—not as bad as they might have had because they were not in very deep yet. "We started with nothing and had nothing to lose," Montfort grins. Worse times came as the twenties went out. "Nineteen thirty was my worst year," he says. "If I had been my banker, I wouldn't have loaned me any more money."

Things picked up in a few years, and Montfort was ready. He turned out the beef, expanded, turned out more beef, and expanded some more. During the war, when feed was hard to get, he built a nineteen-unit concrete elevator in front of the old white one he had been getting along with. The new one holds 26,000 bushels. It stands like a fortress in the midst of a tangle of houses, barns, pens, chutes, sheds, and garages, and the whole establishment, about two miles outside the limits of Greeley, looms on the horizon like a small town in the wheat belt of Kansas.

Montfort is well liked by his men, pays well, and provides good working conditions; but you wouldn't expect it of him after a casual meeting. He is a stooped, dour, stand-offish in-

24

dividual, at least with strangers, though his reserve may melt at meetings of the famous T-Bone Club in Greeley, of which he is a prominent member. He doesn't like much to talk about his business, and you almost have to set a trap to catch him. If you do succeed in cornering him for a few minutes, it takes a corkscrew to get any information at first. After a while, as if a piece of rusty machinery had turned a cog, he will begin to volunteer a few facts without being browbeaten with questions, and you begin to decide that he has a heavy crust on the outside as a result of the pressure he is under and the stupidities and impertinences he has to endure from the curious public, but is essentially kind at heart.

His business is full of uncertainty and worry. He has to buy cattle at the prevailing rate and hope that the price will hold until he can get the stuff off to market. It is not sold, either, until it gets to the stockyards. In these ticklish times the feeder's long-range gamble can produce insomnia and ulcers quicker than any other business known to man.

Still, feeding is a fascinating and exciting way to make a living, and that is one thing Montfort *will* talk about. He can show you the great feeder areas on the map and tell you how they differ and compare. "Cattle feeding in eastern Colorado really came in with the beet industry," he explains. "The beet pulp made good feed, and they brought in the cows—so many of them that they ran out of pulp. So they started a different kind of feeding program.

"This pulp feeding started around 1910. Grain feeding got into high gear after 1930. It is a year-round program. We chop ice out of the tanks sometimes, but never stop. We buy cattle all through the year, but mostly in the fall. There are a good many feeders around Greeley. The Cattle Feeders Club met last night and the men there, in the aggregate, feed sixty to seventy thousand head a year. Bill Farr and his father were there. So were Frank Davis, Bert Avery, the Garrisons, Noble Love, Frank Eckhart, Mage Harrison, Bill Graef, and a lot more.

"Markets? Well, in order of importance to us, they are Chicago, Denver, Omaha, and Kansas City. We truck stock to Denver and move the rest by rail."

You can watch Montfort and Reynolds loading stock any afternoon on the siding a mile or so from the house. They can fill three cars in about fifteen minutes. The pens are crowded with finished steers, every one a butcher's dream with a magnificent rump and smooth sleek sides, Montfort's mixture still staining them a delicate green at both ends. The cattle cars are pulled up to the chute, the bridge from car to chute goes down, and three men send the steers charging and heaving up the incline. Reynolds carries a hickory cane with which he pounds on the planks, whooping and hollering, to speed things up. His roustabout in the pen doesn't even have a whip—just a hank of baling wire which he swings wildly and with gusto. Two or three times he catches a steer by the tail with it and almost has to go along to Chicago. The spooky ones dodge past him, spattering his face with corral mud, but he just grins and takes another swipe at their rears. Every car is filled plumb full and the last steers have to shove hard to get in. It is twenty-eight hours to the first watering stop and they have to stay on their feet all the way, so it is best if there is not any space in the car. They will stand the trip well and lose only 4 per cent of their weight.

And that will bring Mr. Steer to the end of the long trail that started with the birth of a little calf out on the range somewhere. It takes a lot of people to make a beefsteak, and they are all cattlemen of a sort except the last one. Butchers don't wear chaps.

There is one rather elusive kind of cattleman who crops up everywhere in the range country but doesn't get much notice from the cow-country historians. You can call him a trader, though that doesn't set him off very far, since any cattleman is a trader in his heart and soul. Let's say that the standard-model cattleman seems to live to trade, but this type trades to live. His gambling instinct is a little better developed, his disposition is a little more restless, his imagination and his intuition are a little more active than the average. So he buys and buys and sells and sells.

It seems to be a matter of instinct and temperament. Over at Meridian, Texas, they say that Red Nichols started trading jackknives and chickens when he was about seven years old, and he has since developed an amazingly effective technique. He

26

has a frank and open countenance under his red hair and freckles, and a most disarming manner. He never seems to know what anything is worth, but gives the appearance of fretting and worrying over it a great deal. Finally he will say, "I think I can let you have this lot for such and such an amount. I think it will be a good trade for you." He appears to be entirely concerned for the other man's welfare—and maybe he is. Anyway his wife says he takes his trades to bed with him. He turns over on his right side and mumbles, "I think I shouldn't have made that deal." Then he rolls over on the left side and adds, "Well, maybe it was all right."

In 1934 he was farming and raising cows when the bottom fell out of everything. He went to the banker and told him every head he had wouldn't sell for enough to pay off his loan. He wanted to have his loan extended, keep the cows, feed his grain to them, and hope for a break in prices.

The banker was rough with him. "You sell those cows," he ordered.

"All right," said Red. "They're in the barn and if you want them sold, you go sell them yourself. You can pay for the handling, too, and you'll lose more money."

There was some pulling and hauling, but the upshot was that Red began to feed; and when he figured the time was ripe, he took his little herd farther west and sold them for the highest prices brought by any cattle in the county—for enough to pay off everything he owed.

Something like this is the pattern of your rancher-trader—who is not a commission man or broker, but a ranchman who keeps turning his herds over. Sometimes he pastures them or feeds them on his own property. Sometimes he just holds them until he can sell at a profit. Sometimes he ships them out and puts them in leased pastures or feed lots far from his home range. However he manages it, he is a colorful individual with a good deal of the old careless glamour about him.

One of the best traders of them all is Worth Evans, who calls Fort Davis, Texas, his home, but is generally pushing along in his Buick around ninety miles an hour somewhere between California and Fort Worth. Evans is a tall, slender, cowpuncher type with a conspicuous absence of hips. His pleasant, quiet smile does not disturb the impassiveness of the rest of his face. He looks like Owen Wister's Virginian at forty-six—the same courteous reserve, the same feeling of toughness at the core. He asks no odds of anybody, and at the same time he feels no need to convince the rest of the world that he is around.

On the way out to his ranch near Ryan siding he told me how it happened that he is a trader instead of something else. His early history is typical West-Texas-ranch-boy pattern. Born in San Saba County, he migrated with his parents to cheaper land near Sonora, grew up, worked cattle, married a daughter of J. H. Espey. Espey was an old-time cattleman of considerable property and tremendous cattle sense, but Worth Evans did not marry with the intention of standing on J. H. Espey's feet. In 1933, however, he was cleaned out, as was almost everybody else in those parts, and he was glad when Mr. Espey suggested that he had a few sections up in the Davis Mountains which were not burned up, and that if Evans could borrow money to buy cows, he would lease him the pasture land on his note. It worked out very well that way, and when the emergency was over, Worth Evans, his father-in-law, and a couple of their friends formed a corporation. The object was to buy up cows for feeding and speculation. They all have private herds and ranches now (Evans has three), but the company business is kept separate. Evans is on the road much of the time buying stock—any kind of stock. "I will buy anything I can sell," he says. "We ship out by train to Kansas or California and put the cattle on feed. I have even done some business down in Sonora—my cowpen Spanish is pretty fair. In California a lot of the feeding is on beet pulp from the sugar industry. The pulp is put into huge pits and ladled out with a steam shovel. It smells like hell. They also feed miscellaneous stuff—such as raisins—and flavor it up with molasses.

"We lose sometimes. This year we bought stuff in Kansas at thirty-one cents, shipped it to California, and watched it go down to twenty-three. Mr. Espey is the one in charge out there, and right away when our stock hit bottom, he started buying more and putting it into the feed lots—saved some that way. Mr. Espey is a good gambler. When he is losing, he wants to get

28

rid of the stuff quick and start something else. He is the boss and we profit by his judgment.

"We have to expect to lose sometimes. If a man can't afford to lose, he had better stay out of the game."

There you have some of the variations on the traditional pattern of the cattleman—some of the reasons why it is hard to find the real old-time unadulterated article any more. But there are still thousands of Americans who support themselves and their families by turning grass into beef. There are even some areas in the range country where things are about the same as they were seventy-five years ago—where the dress, the manners, the ideals, and the methods of trail-driving days are still maintained. In North Texas and the Panhandle, in spite of the incursions of the wheat farmer and cotton raiser, you still find historic spreads like the Pitchforks, the Matadors, the J. A., and the Three D. The Flying D still operates in southwestern Montana. Angus Kennedy and his co-lessees still ranch as they used to on the Big Lease in the Fort Berthold Indian Reservation in North Dakota. Most of the ranchers in the Nebraska Sandhills are bona fide cattlemen with no fingers in the banker's, farmer's, or anybody else's pie.

It is not just a question of finding a big company or corporation ranch, either. According to the records of the American National Livestock Association there are several hundred small producers for every big one. "Membership records of the various associations show that the average operator runs less than 150 head of cattle."[3]

Mostly it is a matter of topography. Like the Indian, the rancher has always had to back off when he occupied property that somebody else wanted. Sometimes he fought, but eventually he moved. Nesters, homesteaders, plow men, squatters, and farmers have surrounded him time and again—and time and again he has trailed out and started over in newer, rougher country. As a result of eighty years of this process the rancher and the Indian, by and large, live on the least productive land in the nation, and are always trying to keep from losing that. Grass lands in semiarid regions, broken sections where the plow will

[3] *Meat on the Nation's Table*, 3.

29

never go, or mountain slopes too steep for cultivation are the natural habitat of the beef raiser; and fortunately for our supply of steak, there seems to be no doubt that millions of acres of our soil will never produce any crop but grass. It is estimated that around 60 per cent of Texas will always be cattle range.

Look for broken, inhospitable country—follow the trails marked "primitive road" on the map—and there you will find the cattleman. Off in the breaks of the Canadian or the Cimarron he will be rounding up his herd a thousand years from now. Far up under the "quakers" of the Sangre de Cristo he will find his summer pasture (unless the Forest Service runs him off). In the weird and fantastic wilderness of the Bad Lands along the Yellowstone and the Missouri his cows will take refuge from the blasts of winter. Among the spectral forms of the saguaro in the Arizona desert his steers and heifers will find a little grass here and there to put red meat on their ribs. While the climate and topography of our land remain as they are now, this breed of American, with his special history, tradition, dress, and ideals will be with us. There will always be a cattleman.

T 3. The Uniform

HE CATTLEMAN feels himself somewhat apart from the rest of the human race, and therefore he wears a uniform.

There is nothing absolutely G. I. about it—he alters, combines, decorates, and wears his garments as he pleases, but he insists on certain minimum requirements. The height of his heel varies, but the heel has to be on a boot. The width of his hatbrim may range anywhere between three and five inches, but the hat has to be a Stetson or something that looks like one. His pants may be pinks or levis or gray-striped dress-up trousers, but they have to look a little like riding breeches. His shirt may be a Hollywood sport model or blue denim to match his Lee riders, but it has to have the flavor of the outdoors about it. Whether he wears his pants inside or outside of his boots, how he shapes his hatbrim, whether he uses a windbreaker or a leather jacket—

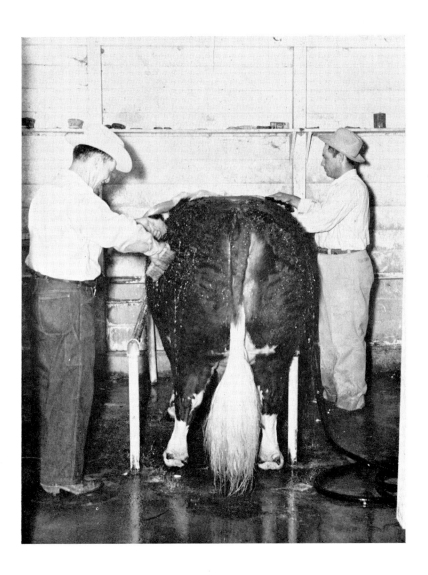

REAR VIEW OF FIFTY THOUSAND DOLLARS
Claude Heard's Beau Zento, Beeville, Texas

those matters are his own business, but he does not get too far from the basic pattern.

He never thinks why he does it, but the reasons are as obvious as the sunset shades of his weatherbeaten hide. A uniform is one of the most ancient and effective of human inventions. First of all, it is useful for identification. You can tell at a glance who is on your side, and who is not. There is a good deal of clannishness about cattlemen, and their clothes help them to get together without too much time wasted. The Stetson hat is as much a trade-mark as the spinach on MacArthur's cap. If you see a pair of boots on one end of a man in a hotel lobby, you can be reasonably sure what the other end of him will look like. And at any distance under a mile the limber-kneed half-hobble of the cattleman's gait gives him away.

Furthermore, there is nothing like a uniform to build responsibility, loyalty, and a certain kind of pride. It is a symbol of strength and virtue. It is something to live up to. It is to some degree sacred. In the army a private on a binge does not disgrace himself. He disgraces the uniform. And since his raiment stands for something he is proud of, he keeps those buttons shining. The high polish on a cattleman's best boots is an expression of the same sort of pride.

That a uniform has as much effect on the lesser breeds as on the chosen people is another obvious fact. There has always been "something about a soldier," but many a war bride has discovered that the glamour can come off with the clothes. There is something special, likewise, about a nurse in uniform, a policeman, or a richly bedizened member of the Exalted Order of Prairie Dogs. One reason nuns and Christian brothers make such good teachers is because of the extra leverage given them by their special dress. A Texan in boots and big hat walking down Fifth Avenue always gets attention from the callous crowd, not because he looks outlandish, but because most of the second-lookers know what his clothes stand for.

The rest of the country knows it, too, and wants to get in on the benefits. The first thing a tenderfoot does when he leaves the train at Tucson or Laramie or Billings is to attempt to disguise himself as a native. The millionaire oilman who buys a ranch as a hobby immediately gets himself some glittering boots and joins the Cattleman's Association. In towns where Annual Rodeos, Stampedes, or Frontier Days celebrations are part of the civic program, everybody has to impersonate a cattleman—salesmen, street sweepers, retired ministers, and all. The last time Truth or Consequences, New Mexico, entertained the cattle growers of the state at a quarterly meeting, the city fathers had "bull pens" set up at strategic points for the retention of rebellious or negligent citizens who appeared on the streets without some decoration. The acceptable minimum was a red bandanna around the neck. Anybody who got caught and penned had to pay off and make up his deficiencies at once. They do it in El Paso and Las Cruces and Brownsville and a hundred other places, to the complete disgust of the old-timers.

"The cowboy is dead," says one disillusioned writer. "There are, to be sure, an assortment of odd characters masquerading in the shadow of the glamour which once was his. Men will ride all day in a pair of so-called riding boots. In fact, riding boots—which do not even remotely resemble those worn by the old-timers—are becoming a fad. The taint of cheap and gaudy commercialism grows stronger when the Chamber of Commerce induces bald and paunchy business men to go tottering around on a pair of high spiked heels to advertise rodeos or to entertain tourists. The tourists are undoubtedly entertained and the C. of C. excuses its actions on the grounds that it only adopted the practice as a last resort upon discovering that few modern men are virile enough to raise a decent set of whiskers. . . . All of which puts the horse before, behind, and in the cart, all at the same time."[1]

Sometimes a group of outraged citizens decide that this masquerading has got to stop, and they take measures to see that nobody wears the uniform without proper qualifications. They act in about the same spirit as a West Pointer when he catches somebody impersonating an officer. At Madisonville, Texas, the Madisonville Sidewalk Cattleman's Association has

[1] Toi Kerttula, "In Memoriam," *The American Cattle Producer*, April, 1949, p. 12.

sprung up to defend the breastworks against invasion. No one can wear boots in Madison County unless he owns cattle. Offenders when detected are dunked in one of the two horse troughs on the shady courthouse square.

Something may be made of the fact that the president of the organization is a local physician—Dr. J. P. Heath.

A uniform can, of course, be useful in purely practical ways, and the cattleman's regimentals have always been regarded as the last word in adaptation to specific needs. The high heels keep the foot from slipping through the stirrup and are useful as brakes when a cowboy roping on foot leans back against the lariat. Chaps are for protection and for warmth. The handkerchief keeps dust out of the nose and ties the hat down over the ears in a snow storm. The sombrero provides shade, sheds water, makes a drinking cup, is useful in signaling, doubles for a pillow, and stimulates a bucking horse. The Mexican *vaquero*, they say, worked these things out, and the American cowboy simply took over the Mexican's outfit along with his land.

All this was true in 1880. But the picture needs some retouching in 1950. Some items in the cattleman's wardrobe are still the best available for his job and his environment. Others he wears because he wants to and not because he has to. The big hat and the chaps are really useful, but the boots, for instance, are not as indispensable as they once were. On small spreads there is not much riding to be done any more. Many operators do not want any yipping and galloping around their expensive stock, and the boys work the cattle mostly on foot. More and more you see them in heavy work shoes clumping around the corrals, although they normally appear at the Saturday-night dance in high heels and narrow toes. Like soldiers, they wear old clothes for the dirty jobs, but get back into uniform when they are on parade.

Some cattlemen try to eat their boots and have them too. Bootmakers turn out medium and low heels as well as the traditional type. It is generally thought that the low-heeled boot is the direct result of the rodeo. Calf ropers can't risk turning an ankle when they come off that horse. Regular cattlemen seem to think the lower heels are worth borrowing, however, and some of them will come right out and admit that what they would like to have is something that looks like a boot and feels like a shoe.

H. H. Matthews does most of his work managing the Elsinore Ranch in a jeep. "I don't ride much except in the busy season," he says, "and I don't need those high heels. I keep trying to get the bootmaker at Fort Stockton to make me a shoe with a boot top on it, but he won't do it—says he doesn't want to change the last. He made me one pair that was about right, but the toe was still too narrow."

On the other side of Fort Stockton lives Farris Baker, who doesn't hold with these ideas at all. The boot is actually more comfortable than a shoe, he thinks. "It supports the arch."

This is a touchy subject, but an outsider can hardly help feeling that the real reason for the persistence of the boot is the fact that it is part of the uniform and therefore sacred. How else can one explain the tender concern lavished on his footgear by every cattleman who can afford to pamper himself? Cotton Whitehead pays seventy-five dollars for his boots, and they are works of art, elegantly shaped and shined until you could see to shave yourself in them. Most ranchers pay between forty and sixty dollars for their dress-up boots and have them made by hand if the old bootmaker is still able to work. Otherwise they buy Hyer or Western or some other standard custom brand, having their foot measurements sent in to the factory.

One of the last of the old bootmaking families is still at work in the person of D. V. Ross, who set up a new shop just off the main street in Billings, Montana, early in 1949. His father was a top bootmaker in Amarillo, Texas, and his mother and brother are in the business there now. All his kinfolk, he says, have followed the trade. In 1948 he decided to join the other Texans whose exodus to northern regions is one of the striking developments of recent range history, and smelled a good opportunity in Billings. The Maverick boot shop is his only competition and his pocket-sized establishment is doing plenty of business. Any time you drop in, there will be at least one connoisseur of leather hanging around to talk shop, and D. V. will probably be just emerging from his work space at the back of the store—a small, boyish, enthusiastic fellow, begrimed but smiling.

When I was there, a husky citizen who answered to the name of "Doc" was present with his wife, both of them full of faith in Ross boots. Doc said he had ordered his first pair from D. V.'s father back in the Panhandle many years ago, and was still wearing some made four years before by D. V. with tops taken off an older pair created by Father Ross himself.

"The family has always turned out good work," D. V. says modestly. "My father was the best there was. But things change even in our business. Heels are wider and flatter than they used to be. Tops are going up. People want them up to fourteen inches now. And we put on more decoration than we used to. My trouble is that I have to educate my trade. Lots of cowboys don't know the difference any more between handmade and factory-made boots. They can buy factory boots for around thirty dollars. Hand-made boots come a little higher—as low as $42.50, but usually more. Fifty dollars is about average.

"Business is good, though, and I have a wonderful opportunity here. If some of the fellows in Texas knew about it, they'd move in right now."

The hat is as important, and just as expensive, as the cattleman's boots. The Stetson is still standard. Quality is indicated by the number of x's on the sweatband, and you can make some deductions about how well a man has been doing by looking inside his hat. Seven x's mean that he has the money to buy the best. If he likes a five-inch brim, the best costs ninety dollars.

This applies only to the ranchman's convention or Sunday apparel. For work he generally appears in the worst old weather-beaten felt you ever saw, sweat-stained, shapeless, and dirty. In summer he takes to a straw hat of no particular pride or ancestry, though it still has a distinction of its own. It costs him about $1.58, but he puts it in the bathtub when he gets it home and shapes it to suit him—usually with the front of the brim curled as if he took it off and put it on with both hands. For summer dress wear he sometimes adopts a "Bulldogger" Panama, which looks like a Stetson in straw.

There was a time when you could make a guess about where a cattleman was from by looking at his hat, according to Philip Ashton Rollins:

In the Southwest, the crown was left at its full height, but its circumference above the summit of the wearer's head was contracted by three or, more commonly, four, vertical, equidistant dents, the whole resembling a mountain from whose sharp peak descended three or four deep gullies. In the Northwest, the crown was left flat on top, but was so far telescoped by a pleat as to remain but approximately two and a half inches high.

Few men of either section creased their hats in the manner of the other. A denizen of the Northwest appearing in a high-crowned hat was supposed to be putting on airs, and was subject openly to be accused of "chucking the Rio," vernacular for affecting the manners of the Southwesterners, whose dominant river was the Rio Grande. Present-day Northwesterners, faithless to this tradition, have foresworn the low crown and assumed the peak.[2]

Rollins was writing in 1922, and things have changed considerably since. The vast variety of crowns, brims, and creases makes any generalization dangerous, but it does seem as if the brims get wider and the crowns get higher as you go north. The North Dakota cattleman, as I have observed him, seems to be holding on to the hat styles of the Texans who pioneered in his region, while the Texans in Texas don't despise a sand-colored Stetson with a three-inch snap brim.

Boots and hats are the staples of the cattleman's uniform, but the accessories are always interesting, and always changing in one way or another. Take a thing like spurs. Spur making is an industry with its heroes as legendary in their own fashion as the Edisons and Fords in other types of manufacturing. Ellsworth Collings, historian of the 101 Ranch, likes to talk about J. R. McChesney who invented the "gal leg" spur. "Before McChesney's time," says Collings, "spur making was usually intrusted to the ranch blacksmith, who turned out branding irons, windmill parts, and wagon jobs as well. But cowboys liked women—treated them with great respect, but liked them. And when McChesney invented a spur with a shank shaped like a girl's leg, they went for it. McChesney added to the value of his product by perfecting a one-piece spur. Before his day the shank was riveted on and could come loose."

[2] *The Cowboy* (New York, Charles Scribner's Sons, 1922), 103–104.

As business increased, McChesney moved to Gainesville, which was a minor cowboy capital in the early days, and finally to Pauls Valley, Oklahoma. There he had a two-story building and employed twenty or thirty spur makers. Then came the first world war, and he got the contract for furnishing spurs to the United States Cavalry. It was a big contract, and might have made him a tidy sum if the price of steel and labor hadn't skyrocketed. Because he had not anticipated these developments, McChesney found himself in trouble, but he did not run out on his contract. He borrowed money—got deeply into debt and grieved over it—and finally died. He is still remembered with respect and admiration as a pioneer in cowboy crafts.

P. M. Kelly of El Paso, Texas, carries on in his footsteps and enjoys top rank today.

Saddles are another special subject that cattlemen can talk about by the hour. Great saddlemakers, like "Tio Sam" Myres of El Paso, are still to be found, who turn out jeweled creations worth two or three thousand dollars for display, for prizes, or for fun. But even in their business things do change, and if you think a saddle is just a saddle, ask one of them what is going on. Henry H. Schweitzer of Matador, Texas, who served his apprenticeship under Tio Sam before Mr. Myres left Midland for El Paso, knows all about it. He has been working for Matador cowboys and their neighbors since 1924 and has seen saddle fashions come and go.

"The shape of the saddle is one thing that has changed," he says. "We have a lower cantle now, and a lower and sometimes bigger horn. The front part of the saddle seat has been lowered and distributed where it might hurt you. We don't see silver horns any more. Square skirts seem to be coming back. We used to see nothing but square skirts in the early days. Then everybody wanted round skirts—and now a few people are asking for the square ones again. People want their saddles comfortable, too. I make quite a few with quilted seats, and we're just beginning to use sponge rubber. Get up on there and see how you like this one."

I got up on the display saddle, which had a quilted seat, and felt a caressing softness unlike the bottom-battering rigidity of the saddles I used to know.

"What would Shanghai Pierce and Colonel Coffee and all the old-timers think about a contraption like this?" I thought, as I climbed hastily down. "And sponge rubber! . . . We'll have Beautyrest saddles next!"

The saddle used to be almost a part of the uniform. Even a horseless cowboy owned his saddle, and no ranch owner ever thought of providing "hulls" for his hands. But there, too, times are changing. Jesse Pearce of Catarina, Texas, has half a dozen saddles in his harness room for employees who bring nothing but themselves. Earl Monahan, however, says they had better not come around to work for him without at least that much equipment. "When I have to provide saddles for my cowboys," he threatens, "I'll get out of the business."

As for smaller items in the traditional costume, some of them have gone with the years, too. One hardly ever sees a cattleman with an old unbuttoned vest on any more, in spite of the tradition that a cowpoke would throw away a suit and keep the vest just so he would have a place to put his Bull Durham. Gloves come out in cold weather and stay in the rest of the time, except in the movies. Chaps are worn generally all over the range, even when they don't seem to be necessary. In brushy country they are lifesavers, of course, but some of the cowboys will wear them at corral brandings for protection and (on cold days) for warmth. Rodeo cowboys wear them, partly, at least, because they are expected to—the uniform again.

To know what to wear, or what cattlemen are wearing, the best thing to do is to stop at one of those capacious stores that flourish in every big town in the cattle country and in many of the little ones. They sell everything for the cowboy, including the rodeo and Hollywood varieties, sometimes handle a line of regular haberdashery, and occasionally add curios and Western craft work for the sake of the tourists. You become aware of such a place two hundred miles before you get close to it in northern Colorado. About every third turn of the road you see a reminder that F. M. Light and Sons of Steamboat Springs are selling hats, levis, and everything else a cowboy might need, except possibly a truss. On the back of the signs the firm has lettered an inscription advising you to turn back if you have missed this wonder of the West. By the time you get to Steamboat, your curiosity will probably prod you into a visit, and if it does, you won't be sorry. You will be greeted at the door by C. W. Light, probably —a big, happy man who loves to meet folks and bubbles with information about his business. He has been there since 1905 and he knows what the trade wants. The store sells custom boots with mostly medium heels and tops—no walking heels yet, though heels are getting lower, and no really high-topped boots.

Shirts must not be too fancy—denim for work, wool gabardine for dress. At the moment everybody has to have pearl snaps instead of buttons. "The boys may not want them next year," says Mr. Light, "but now they have to have them, even on denim shirts. Stetson is the most popular hat. A cowboy will always buy one unless he is hard up. Thirty years ago he asked for a four- to five-and-one-half-inch brim. Now he wants it from three to three and one-half. Yes, we sell a few saddles with sponge-rubber seats, but we don't stock them. Older men wear bandannas, but the custom is going out among the young ones. Sometimes a boy uses a silk handkerchief for a necktie, but the demand is mostly for bow ties made up with an elastic band. Our spurs are all steel. We tried aluminum, but it marked up easily and rode up on the heel. Aluminum bits are standard, though.

"About the only thing we sell as much of as we did forty years ago is the old Fish Brand slicker. We call them pommel slickers. They cover the whole saddle. Yes, business is pretty good. We have six men in the store and on the road."

Thanks to the cattleman's pride in his tradition, the envious imitation of more pedestrian Americans, and the advertising of Mr. Light and his colleagues, ranch costume is gaining popularity. And this results in some interesting complications—such as the way some of these indoor cowboys can fool you. One of the most authentic punchers I ever saw wandered into the Wishing Well store and curio shop two miles out of Paulden, Arizona, one day when I was quizzing Mr. and Mrs. Adams, the proprietors, about Easterners who had bought up local ranch property. He was tall, lean, and muscular—young and rather fresh looking with a noble beak of a nose and a rosy complexion. His outfit was right—dusty boots, spurs, jeans, red checked shirt, hat just battered enough and just enough curled in the brim. He was apparently only an extra-clean-cut, extra-quiet local cowhand till he let slip a broad "a" and upon questioning admitted that he was from London. By the time I had extracted this crumb of information I was beginning to detect a slight chill of British reserve which said, "Thus far shalt thou question, and no farther."

When he rode away without much leave-taking, on a big black horse, I noted the ramrod spine and military seat in the saddle, and was not too much surprised when the proprietor and his wife whispered excitedly that this was an English nobleman visiting with his wife at the Bud Lighten ranch near by—a man whom Lighten's movie-producing activities had turned up. He was taking refuge from diplomatic duties in Washington on the Lighten ranch and maintaining an unnecessary incognito lest he should be bothered by too many common Americans. The point is that in spite of his glacial interior, he had succeeded in getting himself to look like a warm-hearted Westerner. You can't always tell by the outside.

This fact has made some cattlemen a little shy about wearing the uniform lest they be classed with the drug-store phonies. A San Angelo woman once said to Mrs. Arthur Harral, "I'm just beginning to realize that the people who wear boots and big hats are not the ranchers."

Cattlemen themselves are apt to look for additional evidence before greeting a fellow traveler as a member of the fraternity. Clyde Buffington says it is possible to tell by a man's knuckles. If he has done enough roping and flanking, they develop into miniature door knobs. Another method is to go look in one of the favorite haunts where they gather to dicker, drink coffee, and exchange range gossip. Sometimes it is a hotel—not the newest or the biggest, usually, but the one where their fathers hung out. Sometimes it is a coffee shop, or the Elks Club, or even a barbershop. Whatever it is, that is the place to start looking for the real cowmen.

A genuine old-timer can be told by the height of his boots. In the early days the low-topped boot was unknown. In Wyoming in the nineties, says John K. Rollinson, "all the men were well shod in good-looking riding boots, except the cook, who wore shoes. I learned that the boots were mostly made by a bootmaker named Hyer of Olathe, Kansas, and were generally black in color. All had seventeen-inch tops."[3]

Men who grew up in that era still like their tops pretty high —fourteen inches or more. George Snodgrass calls the low ones "sand scoops." Claude Higgins says he wouldn't own a pair. Real riding cattlemen wear them out in the open country, but in the brush the low boot is the mark of a tenderfoot. George Snodgrass always wears high boots and always wears his pants tucked in. "I would feel like a nudist without boots," he says, "and I can't see any reason for wearing leather next to you and cloth outside the leather."

He takes the same analytical attitude toward his hat. "Most cowmen always wear Stetsons," he observes. "Mine has never blown off. They shade your eyes, keep the rain from running down your neck, and keep you from being beaten to death with hailstones. They make the best eyeshades in the world—for reading, playing poker, or what have you. That is why cowboys wear them in the house. These high-school kids who go without hats puzzle me. I wonder why they don't protect their brains, if they have any— why they wear slickers in the rain, but no hats. They go out in these convertibles and the rain runs down their necks so they have to sit in it. I'd feel like a baby that needs to be changed."

The most reliable criterion of all is the skin test. You can be mighty suspicious of a man in boots if he has an indoor complexion, or even a beach-style coat of tan. But if his face is the color of a good slice of roast beef, done rare—if his forehead is white above the line of his hatband and a shrieking vermilion below—if the back of his neck is seamed and wrinkled by blizzard and dust storm and broiling sun—if there are wrinkles around his eyes and hair in his ears, then you have a right to believe he is what he looks like. He is undoubtedly a cattleman.

[3] *Pony Trails in Wyoming,* edited by E. A. Brininstool (Caldwell, Idaho, Caxton Printers, 1941), 34.

I 4. The Code

IT WAS QUITE A FIGHT WHILE IT LASTED.
Stanley Jeffers, a thirty-three-year-old river patrolman for the hoof-and-mouth quarantine along the Río Grande, was armed when he got into his car on the afternoon of August 12, 1948. River riders may carry arms only when on duty, but Stanley thought he had good reason for carrying his. Probably it was the same reason which induced him to take along his brother Joe, just in from California, when it was time to go down to the river for water. In the back seat, along with the buckets and empty milk cans, sat Eugene LaFever, who had arrived from California with Joe.

The occasion for these precautions was the Babb family, who lived two blocks away in the little town of Lajitas. W. L. was in the cattle business, running the Lewenthal ranch near by, and Smoky worked with him. There had been bad blood between the Jefferses and the Babbs for some time, and both sides were about ready for a showdown.

The trouble seems to have started with Stanley's handling of his job as river rider. Sam Jeffers stated on the witness stand that the Babb boys resented the killing of stray cattle belonging to their friends in Mexico.[1] Whatever the original cause, the feeling between the two families had become so strong by July of 1948 that W. L. Babb told the Jefferses not to come to his well any more, forcing them to haul their water supply from the river. On August 10, according to his own testimony, W. L. went to Alpine and bought a new pistol. While there he complained to Leo Bishop, a quarantine official, that the Jeffers boys were annoying and endangering him by indiscriminate shooting around their house, and asked to have them transferred. Bishop said he would see what he could do.[2]

On the twelfth, Stanley and Joe were at it again—"shooting blackbirds," they said.[3] Smoky Babb said he thought the bullets

[1] Special dispatch to the El Paso *Times*, August 28, 1949.
[2] El Paso *Times*, September 1, 1949.
[3] El Paso *Times*, August 28, 1949.

42

were missing him a little too close. About the time the two Jeffers brothers and their guest got into the car and headed for the river, W. L. and Smoky Babb entered theirs and started up the street to make a protest.

They met halfway between the two houses. W. L. pulled up, leaped out, and ran over to the other car. As Joe Jeffers told the story, Babb came up waving a pistol and shouting, "What do you mean, letting them boys shoot down toward our house?"[4]

Joe admitted that Stanley had his own pistol out when Babb's bullet caught him in the head and stretched him out dead in his brother's lap.

The next sixty seconds were as lively as any in the history of Lajitas. Joe snatched the gun out of his brother's lifeless hand and fired at Babb, grazing his scalp and knocking him over. In the next breath he was out of the car and the two of them were blazing away at each other. Joe saw that he needed more fire power, however, and ran for the house, dodging bullets all the way. He came back out with a thirty-thirty, and Babb, seeing that he was outgunned, hit for home. He got away with no more damage than a bullet hole through each shoulder. Smoky roared off in the car, zigzagging for dear life, and escaped unhurt.

When the case was called a year later, the Babbs pleaded self-defense and got a hung jury. On February 1, 1950, they stood trial again. The jury retired at 10:50 P. M. and was back at 11:00 sharp with a verdict of "Not Guilty."[5]

Such incidents, sounding like something dreamed up in Hollywood, happen once in a while in the cattle country. They happen also in Chicago and North Carolina and Iowa. The difference is that fanciers of regional history in other sections are willing to let a killing stand on its own feet. A shooting in Iowa would never be connected with the Code of the Corn Belt. But let the Jefferses and the Babbs get into a fuss on what is left of the old cattle range and it is immediately assumed by visiting Easterners that homicide is still habitual, revenge is still rampant, and necktie parties are still natural among the cattle growers of the West. The Code of the Range, complete with vigi-

[4] *Ibid.*
[5] El Paso *Times*, February 2, 1950.

43

lantes, feuds, unwritten laws, and six-shooter smoke, is given full credit for everything.

The cattleman resents this assumption deeply. The West as a whole was never so violent a place as we have been led to believe. Cowboys left their guns in the wagon on roundups, and many an old-timer never even owned one. As for these degenerate times, most of the pistols worn today adorn the thighs of bountifully-bottomed burghers riding in the parade which precedes the annual rodeo. The burghers may be loaded, but the pistols are not.

The Code of the Range was more a matter of manners than of murder, anyway, and it was not so different from the social standards of decent people everywhere. There were ways of doing business, of meeting others, of conducting oneself, which were accepted as correct, just as there always are wherever two or more people have to associate with each other for long periods of time. Writers on this subject seem to imply that the cowboy was somebody special because he had a code which established standards of decency and honesty; as if husbands and housebreakers and humorists didn't have their codes, too.

The Code of the Range, however, has been discussed so solemnly and taken for granted so long that it might be worthwhile to explain what it actually was and how it has changed in seventy-five years.

At the very beginning we have to face the theory that the cattleman made it all up himself. Ramon Adams puts it this way:

> Back in the days when the cowman and his herds made a new frontier, there was no law on the range. Lack of written law made it necessary for him to frame some of his own, thus developing a rule of behavior which became known as the "code of the West." These homespun laws, being merely a gentleman's agreement to certain rules of conduct for survival, were never written into the statutes, but were respected everywhere on the range.[6]

From the very first point in the Code, it is obvious that many of the cowboy's peculiar notions were only variations of the normal attitudes of other people. "One of the first rules," Mr.

[6] "The Cowman's Code of Ethics," The Westerners' *Brand Book*, June, 1949, p. 1.

Adams goes on, echoing Philip Ashton Rollins and others, "is courage."[7] Well, courage is also a primary quality in a lumberjack, a deep-sea fisherman, or anybody else who follows a trade with a risk in it. Bravery, as a matter of course, was necessary in the cattle business, for "one coward endangers the whole group." That is also true of many other occupations. On the range today grit and spunk are respected. A boy who gets back on a horse which has pitched him off is admired more than one who crawls off to the house. A hand who stays up all night punching at snowbanks with a bulldozer during a hard winter is well thought of, too. Human nature works that way, both on and off the range.

Courage, of course, is never just a matter of enduring hardship or getting back on board a salty horse. Inevitably and finally the term connects itself with physical combat. And that brings us back to the Babbs and the Jefferses. Courage in days gone by meant taking your own part, fighting your own battles. Softer citizens in the settlement might call a lawyer and start a damage suit to atone for affront or injury. On the range you handled it yourself. It was part of the Code.

Yes, it was part of the Code, but the cattleman did not have a patent on it. Lawyers call this the custom of self-redress. It flourishes among frontiersmen of all types and has always been resorted to where courts were absent or unreliable. "Folk justice"—which means usually an eye for an eye—is as old as the human race and is a legacy from our ancestors, whose ultimate statute was the law of blood vengeance. It was not invented by the cattleman or any other Westerner, and has not passed with the passing of the Cattle Kingdom. In the South, among other places, a man who rights his own wrongs with a forty-five is not always ostracized or imprisoned even today.

The rest of the Code follows along in the same track. In the past, certain conditions in a cowboy's life and work made certain habits desirable or indispensable, but they did not necessarily establish him as a separate breed.

Cheerfulness was a virtue. It still is, when men work long and hard together. Cowmen had their share of it; but I do not notice any unusual optimism among them nowadays. When the

7 Rollins, *The Cowboy*, 66–67.

children are sick or the railroad doesn't have enough cars at the shipping point, they can be as gloomy over their troubles as anybody else, and probably get as much satisfaction and relief out of expressing their views. An outsider who steps into a group of cattlemen notes more impassive, dour, and worried faces than smiling ones. And why not? Their business is full of grief and unpleasant surprises. When they are doing well, they frequently look more worried than usual because they know the luck is too good to last.

Then there is the cowboy's respect for women, of which the Western writers make so much. A lot of that is romantic nonsense, too. The old-time puncher, who seldom saw a skirt, was probably motivated as much by shyness as by gallantry in his deference to the ladies. Yet somebody is always making capital out of his supposed innocence.

"I thought she was sure pretty," says Will James of a dance-hall girl, "but what struck me the most is that I'd never seen arms and chest so bare, and I know I felt a little warm around the ears when I noticed that she had no skirts on. It was the first woman I'd even seen without skirts and it'd never come to my mind that they had legs."[8]

Now, you know Will James was better informed than that after living around in bunkhouses and cow camps for years. He was just working the folklore about the Code of the Range for all it was worth. Any real cattleman could set him right. Oliver Wallis, of Laramie, says, "The old-time cowboy was most respectful of women as long as they kept their place. If they let down the bars, one of those boys would go the limit."

All this, of course, has nothing to do with the present. The cowboy is rapidly receding into the past, and the present-day ranchman is a man of the world. His manners toward women are as good as those of any other businessman—and probably no better.

His horse, however, is another matter. Nobody will deny that the cattleman—yesterday, today, and forever—admires, respects, and loves his horseflesh. Sometimes he goes off the deep end, now that he is prosperous, for quarter horses and polo

[8] *Lone Cowboy* (New York, Charles Scribner's Sons, 1930), 202.

JAKE RAINES
Who rode for the S.M.S. until he was seventy-five

ponies. And he still does not like to see man's best friend abused or pushed too hard.

An old-time North Dakota cowman says it for all of them: "I don't like for these goddamed dudes to come out and ride hell out of my horses. Christ, they don't realize that a horse is flesh ' and blood, too!"

Loyalty is another thing the old-timer is supposed to have had—loyalty to the brand, to the outfit, to the men he worked with. There, indeed, the cowboy was different from the farmhand and the factory worker. And where a real one can be found today, he is still a rather steady character. On the big ranches you will find top hands who have been around for ten, fifteen, or twenty years. Tony Hazlewood, ranch boss for the Waggoners on the Three D, has been with the outfit since 1912. George Dubry has stayed by Earl Monahan in the Nebraska Sandhills for twelve years. Hubert Dodge has worked for Raymond Mc-Millin in eastern Colorado for thirteen years. And so it goes. But at the same time the constant complaint among small ranchers today is the difficulty of getting reliable or steady help.

If the humble cowhand had to be loyal and dependable, his employer had to be loyal and dependable in a different way. Like other Americans in simpler times, his great pride was in the sacredness of his word. Legal instruments, contracts, and notes he waved aside with sublime scorn—just wouldn't bother with them.

Evetts Haley tells about two old partners out in Arizona who finally came to the end of the road and decided it was time to settle up. They had never kept any records, but they made their division and parted friendly as usual. A neighbor asked one of them how they had managed it. "Oh," he replied, "it all evened up. I smoke a pipe and George wears socks."

This straightforward way of doing business was accompanied in former times by a sort of delicacy about interfering in other people's business. A newcomer was not quizzed about his name, his past, or his present intentions. As long as he did his job, observed the decencies of the time and place, and caused no trouble, he went his own way and was accepted at his own valuation. All the questions an old-timer didn't ask would have filled many volumes. The inquiries he made would have occupied a few lines. He lived in a time when casual gossip might make trouble, so he did not gossip.

A good deal of this reticence, especially with strangers and outsiders, is observable today. Big cattlemen who move in the world of stock shows, national conventions, and high-pressure publicity have learned how to take and use personal quizzing and what they regard as intimate questions. But small stockmen who live fifty miles from town, and some of the big ones, too, are hard to get anything out of.

Take Henry A. Bledsoe of Cheraw, Colorado. Henry belongs to important people. His father is president of the Colorado Stockgrowers Association, a member of the legislature, and a power in the American National Livestock Association. Henry himself is a fast-moving two-hundred-pounder who looks like what the football coaches are hunting for. I asked him if he had a little time. "Well, a very little," he said as we sat down on the well curb.

"How long has your family been in the business?"

"Since 1890."

"Always at this place?"

"No. First in the Texas Panhandle."

"Then where?"

"Twelve years in Colorado Springs."

"Then here?"

"Yes."

"What was the reason you decided to move over this way?"

"It was bigger country."

"How are you divided up between farming and ranching?"

"We have around a hundred sections—all in pasture. But the country is plowed up all around us."

"Haven't I heard that you are pulling out for a new place farther north?"

"Yes. Nebraska Sandhills."

"What will you raise?"

"Whiteface cows."

The discussion went on that way for some time, a sort of conversational dentistry, every fact wrenched out by the roots.

"Look here, Henry," I finally exploded, "you're no help at all. What's the matter? Are you allergic to any sort of publicity?"

"Well, yes," he replied. "I guess we are. We're not that kind of people. Besides"—he grinned a sly grin—"it brings too many salesmen around."

The best way of finding out what has happened to the Code of the Range is to investigate that favorite myth of the Open Spaces—Western Hospitality. Once the cattleman's doors were open to all, whether the cattleman was at home or not. Has that gone with the longhorn, too? The answer, as usual, is yes and no.

The old spirit is well illustrated by the inscription on an ancient "safe" or cabinet which is a cherished possession of Miller Ainsworth at Luling, Texas. In the shiny modernity of his revamped ranch house the dingy old walnut cupboard stands out like a rusty range cattleman in the lobby of a big-city hotel. On the door are painted the words: "Take what you need. Waste nothing."

Until fairly recent times the motto on the old safe could have appeared on the door of every lonely line rider's cabin between the Río Grande and the Canadian border—and on the ranch houses and headquarters buildings too.

It had to be that way. In a country of great distances and few conveniences a man had to stop where he could. A "No Vacancy" sign out in front would not have made any difference. There was no place else to go. And since everybody in the cattle country had to cover considerable territory in the course of a year's work, all the cattlemen were in the same situation. George might have to bunk with Jim today, and Jim might have to ask for a return of the favor tomorrow.

"It was largely a result of people needing help," thinks Royal Brinkerhoff, who still ranches in wild country in Utah. "Sometimes you had to stop with somebody because there was no place else to go. You might have been able to go on, but your horse couldn't. So you were grateful. Now, with good roads and cars people don't need to be helped so much, so they don't feel so much like helping somebody else."

The co-operative roundup was a great force in making the exchange of hospitality habitual. Neighbors had to work together, and they made common cause in their living arrangements, as a matter of course. To refuse to co-operate, or to put a price on this necessary accommodation, was to be ostracized.

Jesse Pearce remembers when Johnny Green, who married into the Taft family and managed the big Taft ranch with headquarters at Catarina, Texas, ran up against this clause in the Code. New to the country, and not used to having fourteen or fifteen people drop in on him unexpectedly, he put up a sign which read: "Meals, 25 cents. Beds, 50 cents." John Blocker, who ran a big spread a little farther west, notified him that at the next gathering the Taft cowboys would have to bring their own grub and *remuda*—equivalent to telling him the Blocker men would not associate with him or his crew.

Matters went from bad to worse. Nobody wanted to work for Green. Nobody liked him. He suffered keenly under the silent criticism and finally went to John Blocker for advice. "What can I do to get things running again?" he asked.

"You can take down that damn sign and apologize to all the white people in the territory for what you have done," Blocker told him.

Ordinarily a welcome in the cattle country was just taken for granted. "You didn't even ask. You just went in and sat down at the table if you happened along at meal time," says Angus Kennedy.

Such things as baking up a batch of biscuits for any bunch of cowhands who dropped by was part of the pattern of life for a ranch woman. And her husband anxiously encouraged her to be cheerful and prompt about it, if by some chance she was new to the customs of the country. When T. J. (Goob) Saunders of Dickinson married a town girl and took her out to a North Dakota ranch, the first thing he cautioned her about was to be sure to ask anybody who came if he had eaten. "Tell him to put up his horse and feed him, and come on in and eat," he said. Mrs. Saunders took him seriously and found out very soon how necessary his admonitions were. "Our place was a sort of headquarters," she remembers. "We used to have as many as twenty come in to be fed and bedded down. Sometimes they were so thick we would take the mattresses off the bed and put them on the floor, and the boys would sleep crosswise on them. I would have to step over them to get to the cookstove in the morning."

To their friends, cattlemen are still hospitable, very much in the old way. Wherever ranch work involves some exchange of services, the branding or haying or riding outfit expects to be fed at noon. And anybody well known to the family will be pressed to make a visit out of a call. But there are not many old-timers left like Tom Jones of Midland, South Dakota. As I left his ranch house down in a creek bottom south of town, he followed me to the door and said with all earnestness: "It is understood here that anybody who comes by is to be taken care of—fed and bedded down if they want to stay."

I can still see Tom Jones's huge body and gentle old face, and I wish now I had stayed with him that night. A few weeks after I saw him, on July 4, 1949, he was killed in an automobile accident on the road from Fort Pierre. Only a few of his kind are still above ground.

The old ways began to change as much as thirty years ago, and there was ample reason. In the first place, entertainment for all comers became less and less essential. "My husband still likes to ask people to stay and eat," says Mrs. Raymond McMillin, "but it isn't really necessary. We are always busy, and it's only ten miles in to Lamar. It is not expected that people should be entertained any more."

Another reason for the change was prolonged and unbearable abuse of the ranchman's generosity. The legend of the hearty welcome for the stranger spread over the land, and thousands of complete strangers were there to take advantage of it. Sometimes the deadbeats were millionaires. I know of cases— and many more than one—of first-class families back East who felt no hesitation in announcing that they were coming out for a couple of months on the ranch. Worse than that, if they happened to have a difficult child who would be benefited by a summer in the open, they would just put him on a train and ship him out to Texas or Colorado or Montana—wherever the victim lived. You would be surprised at the distinguished names which could be mentioned in this connection.

The less distinguished were no less presumptuous. Mrs. Harvey Witwer tells about one man who showed up unannounced at their Colorado ranch and obviously expected to stay. She asked him how he came to know about their place and he told them that he lived next door to a relative of theirs back East.

On the big ranches the chuck wagon was often the great attraction for these parasites. Johnny Stevens says he had a great deal of trouble when he became manager of the Matador in 1940. "You have no idea how many people want to take advantage of a big outfit like ours," says Johnny. "A lot of them liked to come out and eat at the chuck wagon, and even got so they would bring guests out to eat at the company's expense. Finally I published a piece in the local paper asking folks not to do it. It made some people sore, but we slipped around and told the people who had some right to eat with us that they were not the ones meant. One man was particularly mad about it and complained till I got fed up. He stopped when I told him his wife was the one who had started the whole business."

Ordinarily it was the women who suffered most from these inroads. They were the ones who had to do the cooking and entertaining. Eventually it occurred to some of them that the practice could be stopped, or at least made to pay off, if they began charging for their services, and that was how the dude-ranch business was born. After the episode with their relative's neighbor, Harvey and Stowe Witwer, with the complete approval of their wives, started one of the first guest ranches in the country in 1924.

Other ranch women found other methods. Mrs. Arthur Harral remembers how the revulsion began in the Pecos country twenty-five or thirty years ago. It seemed that people always turned up at meal times. The husband would go out and swing into the old routine: "Have you had your dinner? Well, come on in and eat." It was not unheard of for a ranch wife to come home and find people she had never seen before starting a batch of corn bread in her kitchen.

One West Texas woman got so tired of it that she applied a desperate remedy. The minute she saw a vehicle turning in at the gate, she would take off her clothes and go to bed. She wouldn't get up, either. Her embarrassed husband would have to tell the hungry travelers, "I'm sorry. I'd like to ask you to stay, but my wife is sick in bed."

Many a woman wished too late that she had thought of a similar dodge. Jim and Lola Hunsaker used to live in Leslie Canyon (as Mary Kidder Rak told me their story) and had all they could do to keep things going. One night, after a particularly hard day, Lola had got herself ready for bed and was about to blow out the kerosene lamp when she and Jim heard a hail from the gate.

"Oh, dear," said Lola, with a premonition goading her, "not company tonight—not tonight!"

Jim got up and brought back four travelers—four hungry men for whom Lola had to light a fire in the wood stove and prepare a meal. When they were full and content, one of them rubbed his hands and remarked happily, "We sure were in luck. If we hadn't seen your light, we'd have had to camp hungry."

Back in her bedroom two weary hours after she left it, Lola said to Jim, "I wish I'd blown out that light sooner."

One by one the women rebelled. Mrs. Rak herself was driven to it.

"People who barely knew my husband would come out to hunt," she recalls; "people who had waited on him in a store, perhaps. They never brought anything to do for themselves. But sometimes they brought their wives and children. I finally had to put a stop to it."

"How did you do it?" I asked her.

"I did it in two ways. In the first place, I did not ask them to return."

"And what was the other way?"

"I greeted them with formality."

Mrs. Rak is a woman of great personal force and dignity, and probably had better luck applying this technique than most women could hope for.

"Of course," she goes on, "we are still hospitable to those we know, or to anyone who has a reason to come. We ask them to eat and offer them a place to sleep. But we don't like people to take advantage of us. I remember one woman who left a house full of guests and went to town to have a toothache taken care of. She stayed away several days and finally her husband went in and asked her how long she thought she would have to take treatments. 'As long as those people are on the ranch,' she told him."

Even the men have their troubles. Al Favour says he had to draw a line himself a few years ago about uninvited guests on his ranch near Prescott, Arizona.

"We used to keep line camps where anyone could come, use what was there, wash the dishes, leave a note, and move on. But we had to stop it. They took our hay. They took our wood. They even took the stove if we didn't bolt it down. They broke things up. They actually told us we couldn't come into our own camp if they got in ahead of us. 'We got here first,' they would say. Well, you know we couldn't put up with anything like that.

"Anybody is welcome on the ranch today if he behaves himself. He can stay and eat. All we ask is that he respect our property. But we don't run the line camps any more. I suppose hunters have had most to do with the breaking down of the old customs. They scare our cows and break our fences. Still, I think the old generosity is still there. What has gone is what you might call 'uninvited hospitality.' We still have time to sit down and talk. We are still willing to share with our friends—and with strangers too, if they deserve it."

So don't expect to be received with open arms when you drive up to the ranch house. You may be—but don't count on it. More than once I have had to wait almost until supper time to see somebody. I remember one occasion near Linn, Texas. It was dark, and I was fourteen miles from town when we said courteous good-byes, but nobody mentioned eating—least of all me. I have heard the dishes begin to rattle pointedly in a Kansas kitchen at noon as a gentle hint that I had better be on my way, and I have taken the hint promptly. But on the other hand, I have happy memories of Mrs. Fred C. DeBerard and Mrs. Earl Monahan and Mrs. Ole Fisketjon and Mrs. Farrington Carpenter and many more who treated me like an old friend, although they had never seen me before.

The Code

It helps if you have an introduction, of course. And it helps also if you pay your visits to people below the baronial level. The richer a cattleman is, the less he is inclined to be profusely hospitable, and he is not to be blamed for that. So many people are always trying to get something out of him! Naturally he develops a horny outer covering of suspicion and reserve.

Distance has much to do with it, too. People who live a long way from human society are, as Bob McFall puts it, "glad to see somebody come," whereas those who live among friends and neighbors don't know how it feels to be lonesome.

When I called at the King Ranch, the girl behind the counter in the commissary out in front of the big house asked me if I had an appointment.

"No," I said, "but I have a letter of invitation."

"Well, you'd better call and find out if they can see you."

So I called, and was asked to step across the lawn and talk for a few minutes. Mr. and Mrs. Richard Kleberg were perfectly gracious during the twenty minutes they had at their disposal. But as I chatted with them, I was remembering a place five hundred miles away.

Seventy-five miles out of Fort Stockton I had stopped in at the Augustine ranch to see if I was headed right for Sanderson. A pretty girl with short, curly red hair and big eyes set on a slant came to the door and asked me in. Her name was Jane Augustine Young, and I learned that she was just out of Ozona High School and recently married to a boy who was doing seismograph work for an oil company. She liked the ranch, but she was "glad to see somebody come." She fed me coffee and butterscotch pie of her own making, had her dog Brownie do his tricks for me, and even read me a poem she had composed about Brownie's habits and peculiarities.

For a while I thought I was back in the eighties.

PART II Cattle People

R 5. Younguns

USSELL SANFORD is a red-headed seven-year-old boy who lives on Grandpa Archie Sanford's ranch twenty miles west of Alcova, Wyoming. He is not large for his age and has to have the stirrups shortened up to a point not far from the horn of his saddle, but he is out riding at every opportunity.

Every spring Archie Sanford and his sons, who help him run the place, gather two or three hundred cows and calves in the south pasture across the Sweetwater River and drive them forty miles to summer grazing up in the mountains toward Leon. They all get up at four in the morning and saddle up under the stars; then they splash across the belly-deep creek just above the place where it widens out into Pathfinder Lake. They have to cross the cattle here in order to start the drive to Pathfinder Dam, where they re-cross to the south bank before heading up into the hills.

It takes half a day to round up everything and it is nearly noon when the first bunch comes over the final ridge, silhouetted against the horizon like skirmishers ahead of an army, with riders on either flank.

Archie crosses his cattle at a place where the banks are not too high, and where a willow-grown sand bar makes a good take-off point for the thirty feet of deep water beyond. The herd comes up in a restless and unhappy mood, and it takes the hardest kind of riding to hold them against the bank and force them in as fast as there is room in the water. Russell's father, Wayne Sanford, his uncle Stanford, and the two cowboys, Herb Poorman and Ernest Forsberg, dart off in mad dashes after bug-eyed yearlings who bolt out of the herd and run like rabbits. Russ is not ready for hard riding yet, but he is up at the front beside his grandfather, cackling excitedly when something funny happens.

Archie is one man you couldn't miss on account of his bulk, which fills a specially made saddle. He lets the young men do the hard riding, but he is the general who disposes the troops and keeps the maneuvers going. In an old straw hat and a quilted tan vest, he wouldn't get by the costume man in a Western movie; he does not wave his arms and swear; he does not even join in the shrill yipping chorus of his men as they booger the unwilling Herefords into the water. Most of his directing consists in fanning out his right hand in the direction of whatever needs to be done, something like an Italian fruit peddler apologizing for his prices.

It is all deeply interesting to a seven-year-old cowboy, but the climax comes when a poor pitiful baby calf that shouldn't have been there anyway gets swept downstream. All of a sudden there he is, a hundred feet below the sand bar, nose up, swimming bravely but not very successfully. Herb Poorman, the cowboy in black chaps on the black horse, rides out into the stream and tries to get a grip on him with his hands, but the footing is bad, the horse plunges and struggles, and Herb gets nowhere. He tries again.

"Let him out!" Archie hollers. "Let him out! He'll drown if you hold him there!"

Herb pays no attention. He is already loosening his rope. The first throw misses as his horse steps into another hole. The second try fastens around the neck, and he starts for the north bank, the calf trailing behind and throwing spray like a speedboat.

The horse lunges up and finds dry footing, but the show isn't over. The river bank is steep, and the calf hangs against it. Poorman, who has the rope wrapped around his wrist, is jerked out of the saddle and lands on his shoulders with a crash while his mount pitches away across the pasture in a snorting frenzy.

Russ's high, clear cackle sounds above the bawling and shouting and splashing. He laughs at the thought of poor Herb landing solidly on his neck with his feet in the air, upset by a pocket-sized Hereford. He chuckles and giggles until Grandpa Archie tells him to calm down. He does so at once.

By now the last dripping steer has heaved himself up the bank and trotted off, blowing and shaking his head, to join the herd. Wayne Sanford ambles up and transfers the boy to his own saddle—he won't let him ride the river by himself for a year or two—and the crew cross for the last time, heading diagonally upstream with the water licking at their boottops. In a little while they grow small and far away as they push the herd toward the corrals near the ranch house, and finally disappear around the monstrous granite shoulder of Steamboat Rock.

Russell does not know it, but he has had a fine lesson in being a ranchman. He already takes for granted the early start and the long hours, the movement and bustle, the dust and determination, the lonely hills and limitless sky which make up the cattleman's horizon. He knows the furious pace of the occasional exciting moment, the long, comfortable drag of the hours in between, the good-natured teamwork of the men. He has seen the dead cow, victim of last winter's howling blizzard, and understands that ranch life is not all roping and riding in fine weather. Dinner at two o'clock, taken casually by everybody, reminds him that work comes first, and human beings can think about rest and coffee when the stock is taken care of. Herb Poorman's tumble has demonstrated half a dozen things, including the results of laughing at other people's misfortunes.

Everywhere in the cattle country there are boys like Russ, who will grow up to be cattlemen like their fathers and grandfathers. Third-generation ranchers, carrying on at the old home place, are not at all unusual—probably because a ranch boy begins his training for his future job very shortly after birth.

The ordinary cattleman is a deeply conservative fellow who hates change and dislocation, although he has to expect a lot of it in his business, and what he dreads most is having the ranch go out of the family. He builds up his holdings at the cost of privation, broken bones, failure, and desperate effort. Naturally he cannot bear to think of turning it all over to strangers.

The result is that this horny-handed, leather-bottomed ranchman starts his private propaganda bureau to working as soon as he gets a refill for the family cradle. He turns Manslaughter out to pasture and saddles up old Sadie instead so he can take the baby for a ride. There is a little pair of red boots and a half-pint Stetson waiting when the child shows signs of getting up and hitting a stride. Sometimes the old man's enthusiasm carries him farther than that. At Hayden, Colorado, Francis Miller has all the hay land and irrigated pasture he needs for his two show herds of prize Herefords. But when his grandson was born, he went out and bought another ranch.

How much influence the red boots and the Stetson have on a little boy's mind is hard to say, but it must go deep. The pride of an eight-year-old walking down the street with his father, his outfit an exact duplicate of what his daddy has on, is something to behold. He is wearing the uniform now.

The horse comes next, sometimes simultaneously with the boots. You will never find a boy on a ranch who doesn't have his special horse or pony and his own riding gear. He wants a horse with the same intensity that a town boy wants to drive the family car or that a medieval squire craved the spurs of knighthood. Little by little he becomes master of skills that country boys never consciously learn, absorbing his trade with eager interest, and with the example of the people he most respects as a constant reminder of what he must become.

There is no better way of getting a vocational education.

If he has the right kind of parents, he learns one other lesson at home—the most important of all. "As I look back over the experiences we had in the early days," says Joe Evans, who belongs to an old Texas ranch family, "one thing I can't forget is the teaching of my parents in regard to the kind of men and women we should be. We were taught we should prepare ourselves for life so as to make good husbands and good wives, good neighbors, good citizens, and above all good Christians. It was impressed upon us from the very beginning that we must work —if we didn't work we didn't eat.

"We had no promises from our government to take care of us. We would not have accepted them if we had. Our great ambition was to make an independent living for ourselves and our families and to help others not so fortunate. Instead of wanting our government to do something for us, we wanted to do something for our government. We didn't believe in being guaranteed

Boy Meets Calf
Action at the Houston Fat Stock Show

Courtesy Bill Tipton

from fear and want by our government. The only way we would be free from want was to be afraid we would starve to death if we didn't work."

Joe Evans' economic philosophy may be old fashioned, but the men who have grown up with these ideas, and who sometimes succeed in passing them on to their sons, have built up a pride in their craft, a theory about how life should be lived, and a tradition of keeping the family and its possessions together, which might stand as a model in these uneasy times.

The moment comes, of course, when the ranch and the folks at home are not enough. The boy has to have some schooling. And that is when Father and Mother lie awake nights wondering what to do. If the ranch happens to be near town, or near a school-bus stop, the problem is not serious. But what about the ones who live twenty, fifty, or seventy or more miles from a schoolhouse?

In the old days the problem could be solved by hiring a governess. Some young woman, with or without the formality of a few months at a "normal school," would be engaged to come out to one of the big ranches. She would arrive, excited and a bit fearful, and settle down to teach the children of the family through the winter months. The old custom has almost vanished today. The lure of better pay and wider social contacts draws most young teachers toward community systems. And so many ranch families maintain homes in town that there would be few opportunities for a governess, even if one could be found.

In many places the governess has been replaced by the "ranch school," or "isolated school" as they call it in Montana. The rancher with half a dozen children growing up a long way from town fits up a schoolroom, provides living quarters for a teacher, and the county takes care of the rest. The county superintendent sees that the regular course of study is followed. The school district looks after the teacher's salary and the textbooks.

Russell Sanford is one of two children who spend their school time in a one-room building which Grandpa Archie has fixed up for them. Mrs. Tod Jordan, a pretty, dark-haired young woman with a pleasantly firm manner, lives with the family and carries on as if she were running a full-sized county school. She has a small library in her classroom, keeps in touch with the superintendent's office, and never lets down for a minute.

Any school official in ranch country can point out half a dozen such schools in his territory—Montana, Arizona, New Mexico—it doesn't matter where.

But farther up the leafy canyons in the mountains, farther out across the grassy swells of the prairie, there are families who can't for one reason or another, set up a ranch school. Until recently Mrs. Madeline Irene was an example. She lives in the Freezeout Mountains of Wyoming and might have sent her little boy to the Beermug School near the old post office of Difficulty on the banks of Troublesome Creek (the Beermug School was named for the brand of the old Richards Cattle Company). But Mrs. Irene lives far from any of the three schools in the district, including the one on Troublesome Creek, so for several years she did her own teaching, drawing a regular salary for doing so. In the fall of 1949, however, when her boy was out wrangling the milk cows, his horse fell on him. He was dead when they found him, and Mrs. Irene's school will never re-open.

Another case is that of Mrs. Jeanne McCallister, who lives on the Morgan Route sixty or seventy miles out of Laramie in very rough country. She teaches her six children on week days and brings the mail out from Laramie on Saturdays. Superintendent Helen Irving says that those six little ranch boys and girls, living in one of the most isolated places in the country, are bright as buttons and right up with the rest when they take the state examinations at McFaddin (the nearest community) every spring.

In Colorado the law provides for private instruction when "distance and inaccessibility" prevent children from attending a regular school. The state allows "board and room in place of transportation," which means that mothers like Mrs. Ruby Thompson, who lives away off in the country south of Lamar, get a small salary for teaching their own youngsters. The Thompson children have to spend the last nine weeks of the school year in town, and they take the regular final examinations.

Opha Suckow, superintendent at Glendive, Montana, has

set up correspondence courses for a number of children who cannot get to a school, but this method is not at all common.

On the southern plains, in spite of isolation and distance, schooling does not come quite so hard. Where the roads are good, the country moderately level, and the winters mild, the country schoolhouse is rapidly being abandoned and the districts are undergoing consolidation. Students come in by bus to big plants with auditoriums, athletic teams, dances in the gymnasium, and up-to-date teaching staffs. They enjoy cafeterias, libraries, movies, tennis courts, and luxuries undreamed of in the days of the little red schoolhouse.

Even in the southern regions, however, consolidation has not been made available to everybody, and in such places the hazards of distance are felt almost as much in the country of sun and space as in the land of clouds and cold. At Marfa, on the edge of the Big Bend in Texas, they tell about Mrs. Cherry Bryant, a single-minded woman who places much faith in the value of home life. The Bryant ranch is nearly forty miles from town, but rather than leave home herself or send the children away, she determined to drive them to school and back herself. She rose before daylight, got breakfast, rushed the children through their preparations, unloaded them in town before the last bell rang, and got home in time to fix her husband's lunch. There was leisure for about two good breaths after the dishes were washed, and then it was time to start after the children. When she got home with them the day was finished, and so was she. She moved to Marfa.

Where the school bus comes by in consolidated districts, mothers on outlying ranches have an easier time, but one feels a little sad about the children. Take the case of the Argyle Mc-Allens, who live twelve miles from Linn, Texas, just out of the Lower Río Grande Valley. The bus comes out from Edinburg to pick up children in what the natives say is the biggest school district in the United States. The bus route is fifty miles out—and fifty miles back. Mrs. McAllen loads her two boys (Robert, fourteen, and Jimmy, nine) into the Jeep every morning and takes them half a mile to the corner where the bus stops. They have a thirty-two-mile trip ahead into Edinburg, and it takes them an hour and a half each way. Three hours out of a boy's day gone, just for transportation. With chores to do and homework to get through with, this is a serious amputation. Add it up over a period of years, and a large slice of somebody's life is obviously wasted on the road.

Everywhere in the southern plains and desert country the same problem exists. At Buffalo, Oklahoma, Superintendent Edna Campbell supervises seven consolidated districts with school buses running out to the ranches along the Beaver and the North Canadian. The longest route goes out fifty miles. Parents in the Sopori community in Arizona send their children into Tucson on the Nogales bus—a distance of thirty miles coming and going.

In southern Texas, where the ranches are huge and the distances are more so, some special problems arise—for instance in Starr County on the lower Río Grande. This neighborhood is famous for its flaming politics, patriarchal Mexican families, and determined resistance to American ways. Mrs. Florence Scott, who has charge of the Roma District in the western and northern part of the country, now supervises seven schools instead of the twelve which were operating in 1947. A couple of local schools still hold classes through the fifth grade, but after that everybody rides the bus to Roma.

The population is almost 100 per cent Mexican, and among these conservative people consolidation was achieved only after severe struggle. The parents did not want their children snatched away and thrown with strangers every day—coming home with a lot of radical American ideas. Besides, it was hard on *mamacita* to fix so many lunches every day for her numerous offspring, and poor people could not dress a large family well enough to send them to town school.

Once started, however, the children took over the job of selling the program at home. Mrs. Scott thinks the soft-drink machine did more good than anything else. Joe Guerra, youngest and most active of the great Guerra clan, set up a lunch program which took some of the burden off *mamacita*. And since several years of prosperity have put more money into the family sock, there has been less need to worry about suitable clothes.

One very serious problem still remains. Many of the Mexican families in this district are migratory workers, in a special sense. They leave in August, usually—just before school is supposed to start. One family or more will pull out in a pickup or truck, the back full of wide-eyed children. Sometimes several families will go together. They drift as far east as Mississippi and as far north as Michigan, following the crops, camping out, keeping house in the shacks and hovels provided by their tight-fisted employers, happily living their own lives, singing their own songs, toiling in the fields all day and coming home to the little bit of Mexican Texas enclosed by the light of their fire at night. In November or December they show up again, richer by two or three thousand dollars, sometimes. The father will buy a new car or a new bull for his *rancho,* and feels that life is pretty good. But Juanito and Pepita, entering school two or three months late, are likely to have their troubles.

"It's hard to get them up in arithmetic," says Mrs. Scott. "Their English and general background—and maybe geography —are improved by their travels."

So far, we have been dealing with normal boys and girls with no special talents or special urges. But exceptional children are born to startled ranchmen, just as they are to horseless citizens. What can be done about them?

They have about as much opportunity and encouragement as other children to develop their gifts, though it may take a little more struggle and sacrifice. Earl Monahan, for instance, has two daughters with musical talent, and Earl has always been willing to help them develop it. He has provided a good spinet piano and paid for some lessons. His home, however, is seventy-five miles from Alliance, the nearest point where good instruction is available. Furthermore, the snow gets deep and the wind blows bitter in northwestern Nebraska every winter, and that 150-mile round trip once a week is out of the question for several months of the year. Paderewski himself might have decided to become a veterinarian if he had grown up in Nebraska.

For a different sort of case, take the Bill Hayes family who live twenty-one miles west of Miles City, Montana, above Cottonwood Creek, with cottonwoods and grassy benches below and the smooth slopes of young mountains all around. Bill is a rancher and trader who lives in a rambling, one-story house with a big front porch and a king-sized kitchen. You can sit in the combination living and dining room and see no signs of genius. There is a big, old-fashioned, round dining table in the middle of the linoleum-covered floor, a studio couch, a battered secretary, a tank full of gold fish, sets of unread standard authors in the bookcase alongside the complete and well-thumbed writings of Will James. But if you stay long enough, Mrs. Hayes, a plump and pretty woman in brown slacks with a ribbon around her ringlets of brown hair, will shyly bring out a wad of drawing paper—the complete artistic output of her fifteen-year-old son Dickie. There are crayon and pencil sketches, water colors, and a few paintings, all crude and imitative, but with a definite quality and full of experiment. Much of it is copied from Will James, especially from *Scorpion,* which seems to be a favorite book in the family. But there are cigarette-ad girls, deer, horses, old-time cowboys, and even a gruesome man from Mars. There has been definite improvement in the five years since the first drawing was done, partly the result of instruction from a teacher who is interested in the boy. In some ways Dickie's artistic temperament puts a considerable strain on his parents, but they are hopeful that he can get more training and find out what he can do.

Like most ranch people, they realize that the best thing they raise is children.

W 6. Cowmen in Skirts

HEREVER YOU GO in the ranch country, whenever you call, whatever your errand, your first contact will most likely be with a woman—that is, if you don't count the dogs who come charging out at you, tearing the air to frantic tatters with a hullabaloo that could be menace and could be welcome—neither you nor the dogs are quite sure which.

She may be anywhere between sixteen and ninety, an earnest bride, a flustered housewife of middle age, an energetic grandmother. But the chances are she will be a husky specimen (a ranch is no place for the willowy type), that she will be wrought up to a medium-high pitch by the jobs that are crying to be done, that she will be dressed in comfortable oxfords and a much-laundered house dress, and will be embarrassed because you have caught her with her hair in her eyes. Nine times out of ten she will be as courteous to you as a woman can be when she needs desperately to get dinner going or go to the bus stop after the children or meet the train which is supposed to be bringing her chicks from the hatchery.

If she keeps the screen door between you and her and informs you with an exasperated explosion of breath that she just hasn't a *minute*, you can forgive her—for the ranch would come close to shutting down without her. She is in the position of the man in the control tower at an airfield, or of a very busy lady spider in the center of a very complicated web. All the threads come back to her. She has to send the men out to work fed and contented, and she has to welcome them back hungry and tired. She takes care of the children, the poultry, the sick calves and lambs, the house and the garden. She stands between her men and the promoters, salesmen, tourists, and miscellaneous visitors who find their way to the most remote ranch house. And if there are any community enterprises, charity drives, visits of condolence, school affairs, or church matters to attend to, it always seems as if the men are desperately needed at some remote corner of the ranch and Mother is the one who has to change clothes and go.

There are some old bachelor ranchers who don't know what they are missing, and a few young ones who just haven't completed their arrangements yet, but the average ranch is run by a team of two equal partners—and don't think the woman has the long end of the stick.

It is an arduous life with too many responsibilities and never enough time, but these women would not exchange it for any other. And it does seem to most people who have thought about it that there are fewer frustrated, neurotic, narcissistic, self-pitying, useless females on cattle ranches than anywhere else in the country.

This happy situation is, in part at least, a recent development. Not many years ago a ranchwoman deserved a good share of sympathy if not pity. Those were the days when the wagon went to town twice a year for supplies—and Mother stayed home so there would be somebody to look after things on the place; when she doctored every bunged-up cowboy within a radius of fifty miles, cooked for every stranger who chose to show up at meal time, saw a new dress or a female neighbor about once every two years; when there was about as much chance of getting a doctor or a bottle of medicine as there was of taking a week-end rocket trip to the moon, and that went for the times when the babies arrived, too; when the fuel was wood or cow chips, and she could take her choice whether she chopped the one or gathered the other; when an egg or a vegetable was rare enough to rank as an art object and water was so scarce she washed all the children in the same hatful and then threw it on the roots of the spindly shade tree she was trying to keep alive in the front yard.

Things have changed since those days, and probably the general character of ranch women has changed too. Some of them used to be as rough and tough as the men they lived with. Frank Keogh tells about an old ranchwoman on the streets of Watford City, North Dakota, who observed one of her neighbors leaving town with his purchases on his back. "There goes Joe Jimpson," she said with pity and scorn—"There he goes with a sack of potatoes on his back and I bet he ain't got a drop of whiskey in the house."

Henry Stoes, late of Las Cruces, came over from Austria and started waiting on customers in a general store while the Texas invasion of New Mexico was still going on. I asked him one time about the wife of a well-known frontier character. "Oh," he replied, "she was just an old ranchwoman—she'd spit through a screen door."

Some of the women used to be like that. Belle Starr in Oklahoma and Cattle Kate in Wyoming had their counterparts in other places, ranging all the way from scarlet sisters to real rugged cattle growers who just happened to be women. The old people around Tularosa, New Mexico, still tell tales about a woman known as Broncho Sue who derived her nickname from her talents as a horse trader. She also ran a few cattle, kept herself in husbands in spite of the high mortality rate, and exuded an aroma of vigorous toughness. And she was by no means the only one of her kind.

Far removed from these turgid types was the old-time ranch wife who lived life the hard way but tried as best she could to keep alive some standards of decency and religion in times and places more or less hostile to both.

In that era a ranch girl usually married a ranchman, unless her father happened to be a cattle king, in which case her opportunities for selection were a little wider. If she chanced to be the daughter of a Senator Warren, she might even become the wife of a John J. Pershing. Ordinarily she picked the best available specimen in her neighborhood. Mr. and Mrs. R. N. Everett of Valentine, Texas, lived on adjoining ranches. "All I had to do was drag her across a barb-wire fence," he says. "I'm pretty conservative; I don't like waste motion."

A romance with a neighbor boy was quite satisfactory to a ranch girl, and gave her all the usual thrills and chills. Mrs. Margaret Buttrill, now a widow running a guest ranch near Marathon, Texas, has tender memories of her romance. She was Margaret Simpson, plump, pretty, and fifteen, and she was looking forward to a Fourth of July party. Everybody in the neighborhood was invited as a matter of course. The year was 1897.

Several of her cousins had announced their engagements, and that started one of her uncles to teasing.

"You're getting pretty big now. Won't be long till you'll be thinking about getting married yourself."

"No, I'm not ready yet."

"I'll bet you have the man picked out, anyway."

"Oh, yes, I've got him picked out all right."

"Well, who is he?"

"Lew Buttrill."

They all laughed and laughed. "Why, he's too old for you."

He *was* thirty years old—a good cowman who lived on the

71

YE ranch all by himself. Margaret had never seen him, but she had heard much about him, enough to make her think him different and interesting.

That night he showed up with John Greenwood and John Bates, almost the only other *Americanos* in those parts, and Margaret was pleased with what she saw. He was pleased with Margaret, too, it seemed.

It was a good party. The meal was served outdoors, picnic style, and included all the old-fashioned staples— barbecue, beans, green salad, homemade pickles, pie, ice cream, and lemonade. The ice had to be brought all the way from Alpine. After everybody was too full to straighten up without groaning, they all went inside the house for dancing. The fiddle kept going almost all night. And before morning Lew and Margaret were engaged. On December 23, months before Margaret's sixteenth birthday, they were married, and lived together happily until Lew died in 1933.

Until not so long ago the country schoolhouse was a prime source for wives. The legends about the cowboy and the school teacher are based on solid fact. The shortage of women plus the glamour of a new face did the trick almost every time. An imported teacher who held out for more than two years in a state of celibacy was either completely sold on the independent life or homely as a mud fence.

Nona Condit came out from Iowa to Wyoming in 1890— twenty-seven years old and an absolute greenhorn. Two brothers, already on the ground, had found her a teaching job. At Douglas she had to get off the train and take a stagecoach for Buffalo, with two nights of traveling ahead of her. The rain and the gumbo mud and the chuckholes soon made her seasick, and the future looked black until a traveling salesman from Omaha offered her some "seasick medicine." All she knows about it is that it had an alcoholic base, that it settled her stomach, and that it is a matter of great regret to her that this panacea is now apparently lost to the world.

One Sunday a friend took her to a roundup where she met Al Williams, a big, quiet cowboy with a large nose and a firm face, who was just getting started as an independent cattle raiser. He had been in the country eight years, and thought of himself as a confirmed bachelor. Everybody else considered him so, too, but eleven months after her arrival Nona Condit married him and became the mistress of a one-room log cabin.

Mrs. Williams is close to ninety now, but she still knows what is to be done on the ranch at Banner and sees to it that somebody does it.

Many such women are still about. Mrs. L. C. Brite, loved and respected by everybody in the "Highlands" region around Marfa, Texas, came to teach school in 1894. To cite a more recent case, there is Mrs. Raymond McMillin of Lamar, Colorado. She was teaching in the grades at Granada when she met the secretary of the school board at a dinner party one night. Raymond was and is a determined man. The new teacher was doomed to matrimony from the first and has helped to run the ranch since 1936.

Even in early times a few city women were marrying their way to cattle ranches. The amazing fact is that so many of these matches turned out well. Those women tell about their experiences now with amusement, but sometimes there is the echo of a wail in their laughter.

Mrs. Arthur Harral, small, intense, and full of charming refinements, stepped into the wasteland west of the Pecos direct from Wellesley Hills, Massachusetts. She was visiting relatives in San Antonio when she met a husky, black-haired, slow-talking youth from the cattle and sheep country—a student at Texas A. and M. College at that time.

When Arthur finally persuaded Carolyn Hassoldt to take the plunge and brought her out to the ranch, he dropped her into the midst of a magnificent wasteland seventy-five miles from Fort Stockton, Texas. Her new home was lost among high, rolling limestone hills, dotted with mesquite and cedars, splotched with huisache and sotol, cut by sandy-bottomed arroyos where the impossible roads twisted and writhed until they could charge up a jagged slope in the direction of the next arroyo. It was a good, clean, vigorous barrenness with lots of sky. Carolyn thought she might become reconciled to the loneliness and isolation if she could rearrange things a little more to her taste. The whole outdoors was to be her front yard—no fences! But to dress the house up a little she had a couple of dagger plants dug up and planted on each side of the front steps.

That year (1917) was dreadfully dry. When the cattle got down, they had to be helped up. Arthur had four-horse teams on the road all the time hauling feed, but it was still impossible to pull all the stock through. Always hungry, the poor old cows moaned and bawled around the house, too weak even to make much noise.

The big yuccas did well, however, and soon put out two great red buds which broke into a shower of creamy white. Carolyn was enchanted. Then in the night came an old cow brute, hardly alive. She first ate off the blossoms; then lay down and died right there. "I wept bitter tears," Carolyn admits.

The orphan lamb she brought in to feed and cherish ran up and down the front porch all night long, baa-ing and messing things up.

The lima bean vine she planted to shade the north windows was discovered by a couple of prospecting goats and consumed at once.

After that she allowed a barbed-wire fence to be set up around the yard. But her troubles were just beginning. Her parents came to see how Daughter was taking to pioneer life. They were horrified to find that the family automobile was an old Maxwell touring car with the top removed. Impossible for Carolyn to ride about in such a contraption! Their first act when they got back home was to send a check for the purchase of a new sedan.

"Arthur didn't believe there was such a thing," Carolyn says. "He bought an open Buick, and I shed bitter tears again."

Although he was the kindest and most genial of ranchmen, Arthur needed considerable currying. He asked Carolyn one time if she wanted to take a ride—he had to see a neighbor on business. She would be delighted, of course, and went off to get ready. Dressed in her best, she sat down to wait for him to drive up and help her in. When he failed to appear, she finally went out to the back gate where he kept the car. He was gone.

On his return Arthur found Carolyn cleaning furiously as a means of working off her resentment. Everything was moved out on the porch, and there was no comfortable place inside for any male. Arthur saw that he was far back in the doghouse.

"Why did you go off and leave me after asking me to go along?" Carolyn demanded.

"Well, I had business to do and I couldn't wait."

"But I was ready and waiting for you."

"You were? Then why weren't you in the car?"

Today Carolyn laughs about it. "I broke him of some of that, but I was ready to leave, and my people would have been over-joyed if I had."

"Yes," Arthur chuckles, "she might have, but it was too far to walk to town."

The Harral place is a desert paradise now. The huge yard, seeded to ranch grass, is bordered by a stone wall, with iris planted along the walk to the front gate. Chinaberry trees scatter shade here and there, and what must be the biggest pear tree in existence spreads out over many feet. The house is an old adobe with a tremendous, long gallery in front and room after room budding off to the rear. Oriental rugs, a Baldwin grand piano, good pictures, a big fireplace, and well-filled bookshelves make an oasis out of the living room. Modern bathrooms, electric lights, and butane gas bring the city to that remote patch of rocky country. A stone reservoir one hundred feet in diameter provides stock water by gravity to half a dozen tanks far away in the pastures and makes a swimming pool for hot summer afternoons. The grassy meadow below the house has been turned into a landing field for son Gloster Harral's Stinson Station Wagon.

It would be hard to find another such ranch house where the determination and persistence of one New England woman has imposed a bit of Wellesley Hills on the unyielding primitive-ness of the Pecos country. But Carolyn did it. In her living room you can hear talk about art or books or music. You can get posi-tive reactions on the cultural state of the nation. Draw the cur-tains, pick up the thread of conversation, and Boston is just around the corner.

At the other end of the range lives Harriet Reno, a vivacious little woman whose gray hair doesn't go with her youthful face. You wouldn't believe what she went through, either.

In the early twenties she came out to Sheridan, Wyoming, from her home near Chicago. A visit was all she had in mind, but she met an aloof young man who made her think twice about leaving. Floyd Reno was the son of a prosperous cattleman whose headquarters were in the country sixty-five miles from Gillette and about the same distance from Douglas. He was nineteen. Harriet was eighteen. In due time they were married and went off to live on his homestead.

Father Reno had enough for everybody, but he did not believe in scattering it around. "Little ships must stay close to shore," he said. Floyd and Harriet had $250 to live on during the first year—and they lived on it. While they were building their first log-cabin home, they kept house in a sheep wagon. Range people point to those canvas-covered contraptions as the last word in efficiency, with stove, bed, table, and everything else arranged to fold up, disappear, slide under something, or stand out of the way in a corner.

Harriet is completely disillusioned about them. They may delight the soul of a sheepherder, but they are an exasperation to a home-minded woman.

They had no meat in the summer—no fresh vegetables—and not much of anything else. "My friends think I am crazy when I tell them I was hungry for five years," she says; "but I was."

"About the third year I finally got to town. I bought a sack of tomatoes, took them up to my hotel room, and went through the whole bag with no fixings but a salt cellar and a ravenous appetite. Then I went out and bought four dishes of ice cream, and tapered off with an assortment of hard and soft candy."

For three years the two of them took care of the elder Reno's herds for half the increase. They provided the feed and the labor. Harriet kept up the woman's end, cooking for twenty-two men. That job she stepped into on the day of her arrival, and she had not had any previous experience.

To make it harder, the people on the ranch had to be shown that she was not a misfit and a nuisance. "Ranch communities are pretty democratic. In this country the family and the hands eat together, and they let me know when they didn't like what I did. I used to get so mad I could hardly stand it."

They made two trips a year to town in those days. It took two days to drive in and two to come back, and they camped out along the road. "The outdoor enthusiasts can have all of that sort of thing they want," Harriet says. "I've had enough.

"I tell Floyd I wouldn't go through that again for any man. He always reminds me that he was there too."

Harriet has many conveniences on the ranch now, plus a house in Sheridan where she lives during the winter while the children are in school. She is an eager dancer, a charming conversationalist, and a very pretty woman. Apparently those early struggles did not do any permanent damage, beyond leaving a few scars on the memory.

Not every ranch woman has such painful recollections. There were places where life was much more gracious—for instance, the eastern part of Texas, where the traditions of the Old South took off some of the wire edge.

Maud Wallis Traylor likes to remember how it was when she married into a big clan of Gulf Coast cattle raisers. She was the fifth Mrs. Traylor on the place, but only her husband's mother was known by that title. The daughters and daughters-in-law were Mrs. Will, Mrs. John, Mrs. Champ, and so on.

"Those sisters were spoiled within an inch of their lives," Maud recollects. "They married rich men, kept colonies of Negroes, and never had to turn a hand except to enjoy themselves. The only rough spot was the custom of early rising. Everybody got up and ate breakfast by candlelight so the men could get on the prairie. We were supposed to behave like ladies, of course, and the greatest sin we could commit was riding 'astraddle.' I had a good, comfortable sidesaddle and didn't need to ride like a man, but I sometimes did, just for devilment.

"We had our best times during roundups. All the Traylor women would ride out to the pastures right after sun-up and watch the operations. We were expected to keep out of the way, and of course none of us thought of lending a hand. At noon they would stretch a tarp for us and we would eat the fine food

77

prepared by old 'Ronymus, black as the ace of spades at midnight and ugly as homemade sin, but a wonder at steaks, cream gravy, dried fruit, and biscuits that would make you slap your grandmaw."

A woman's life on a modern ranch is no such picnic as that. Neither is it the purgatory endured by Carolyn Harral and Harriet Reno. I had a better-than-usual chance to see what it is like one March morning when the home demonstration agent at Fort Stockton, Mrs. Grace Moore Martin, took me twenty-eight miles down the Marathon road to see Mrs. H. H. Matthews, who helps her husband manage the Elsinore ranch.

We found her out in her yard, a stocky, vigorous woman with a wholesome, cheerful, sun-cured face and large, capable hands. The whole history of a couple of ranches is worked into those hands. She had on a serviceable but becoming green wool dress, wool-and-rayon stockings, and stout brown oxfords. I noted that her cruising speed was well above the normal rate, and that every now and then she lengthened her stride with a little spring, as if she were about to break into a gallop. I saw the reason for this as she showed us around and talked about her job.

She raises all sorts of garden stuff, takes personal charge of the farm, raises peacocks commercially, runs her house, and keeps up her music. She had just put in a grape arbor, a shed to protect her young rosebushes and carnations, a covered cellar for ranunculus bulbs, a new berry patch, and new strawberry beds. A yard man helps her, but she goes by him as if he were standing still.

Her house is a huge old adobe built and furnished for use. The floors are covered with linoleum or painted and left bare and clean—no luxuries about the place except a fine grand piano in the music room. There are books and magazines and comfortable, ungraceful old chairs in the living room. The kitchen is small but completely modern, with an electric stove, electric refrigerator, and hot and cold water in the sink.

Her inside jobs include cooking three meals a day, canning (three hundred cans last year), and making jolly conversation for neighbors and guests. The day I was there it took her about half an hour to prepare a meal of ham, beans, sweet potatoes,

RANCH GIRL
Donna Glee McOllough of Monte Vista, Colorado,
queen of the Ski-Hi Stampede in 1949

corn bread, and ice-box fruit cake for herself, her husband, her neighbor Mrs. Frank Warren, Mrs. Martin, and me.

I asked her if she got lonely out there by herself. She made a sound that could have been called a snort. "I wish I had two hours to be lonely in. I work fourteen hours a day and wish the days were six hours longer."

Mrs. Warren, young, plump, and pretty, agreed. "There just isn't fifteen minutes to relax on a ranch. Frank and I get up at four-thirty and from then on if it isn't the door bell it's the telephone. Sometimes till midnight."

"Yes," Mrs. Matthews added, "and when the men are working out, we have to have dinner ready to serve from eleven-thirty on, and sometimes they don't come in till four. You'd be surprised how these men hate outdoor cooking. Mr. Matthews doesn't mind it so much if the food is good—and clean—"

"Frank says chuck-wagon food is filthy, and he's pretty near a connoisseur of food, too. He comes in whenever he possibly can. He says he wonders why people will come out to eat chuck-wagon cooking as if it were something special."

We talked about the long drought and the heavy job of feeding the stock—the deep freeze unit that simplifies the meat problem—the Matthews' two sons (one a Ph.D. from the University of Chicago now teaching at Texas A. and M.)—the truckload of steers that disappeared last year and the bunch of sheep that the Warrens were missing at that moment (Warren was out in his airplane looking for them).

Dinner over, Mr. Matthews headed back for the pasture, and Mrs. Martin and I said our good-byes and took leave. Before we were well out of the yard, Mrs. Matthews had dashed off on some pressing errand at a gait which was almost too near a run for respectability.

It might seem to a city dweller that a life like this would lead to nervous breakdowns all over the range, but it actually seems to agree with the women who follow it. They all say they like the ranch because there is something new every minute. Mrs. Evetts Haley loves living off in the breaks of the Canadian River because "if it isn't a sick calf or a broken-down windmill or a cow bogged in a mudhole, it's something else. You never have time to be bored. And at the same time you can enjoy the quiet of the country."

It would be wrong to imply that these women live like lady hermits and never get off the ranch. Their social life can be rather intense, and can get them into all sorts of clubs, circles, auxiliaries, and boards of this and that.

The ranch women near Monte Vista, Colorado, enjoy the Rock Creek Club, which has been disseminating culture, gossip, and good food since 1914. The program committee arranges for speakers, book reviews, dances, and plain old-fashioned talk festivals which culminate in tremendous lunches.

I asked Mrs. Luke McOllough about the last book review. "Well, one of the members reviewed *Cheaper by the Dozen*," she told me, "but I didn't get to hear it. I was on the refreshment committee and was out in the kitchen. I read a condensation of it in the *Ladies' Home Journal*, though. I don't have time to read a whole book."

Those who don't have the Rock Creek Club have something else. Mrs. Ernest Ham helps with the home demonstration program and sponsors 4-H Club work in the vicinity of Piedmont, South Dakota.

Mrs. Raymond McMillin drives ten miles to Lamar, Colorado, to meet the members of the Book Club, Study Club, and Bridge Club.

Mrs. Alfred Collins takes much satisfaction in mothering the little Episcopal church at Crestone, Colorado.

Mrs. Elmer Campbell of Morgan, Texas, is on the board of nationally accredited flower judges and pursues her specialty in Dallas, Fort Worth, and Waco.

At Columbus, Texas, Mrs. Elbert Tait, Mrs. Henry Frnka, and Mrs. Lester Bunge, among others, belong to the Music Club directed by Miss Lillian Reese. They work long hours getting ready for special programs and the big annual concert.

Mrs. Bob Donelson likes to travel. Last summer she and Bob left their Burbank, Oklahoma, ranch, flew to Alaska, and saw the whole country by air. "We could have dropped a rock on Russia," she says.

In the range country of Utah, all activities center in the L.

D. S. church. Mrs. R. J. Brinkerhoff of Bicknell gets around to meetings of the Women's Relief Society and to church services. Bicknell is a small place and too far from any big town for frequent excursions. Church work provides about the only social outlet.

Ranch women nowadays even belong to country clubs, when they are within reach. Kay Crum and her husband, J. Y., drive fourteen miles regularly to attend functions at the club in Weatherford, Texas. Every Friday night there is a family supper and bingo party. Each family brings one dish and they all help themselves to everybody else's dishes. Members give dances at the clubhouse to repay their social obligations and keep up with their friends. The men have a stag party at the club every Monday night.

All this leads to the conclusion that a fair amount of the grief has been taken out of the woman's end of the ranch business. The change appears particularly in the matter of "outside work." Oliver Wallis comments on how it used to be on the ranch in the foothills of the Medicine Bow Range in Wyoming: "Some women worked outside during my early years. My sisters never did, but there were some who had to. During World War I, when help was so short, lots of women took men's places and made good hands, too. They were always perfectly free from embarrassment—unless they didn't want to be. The boys would never intrude on the privacy of a woman unless she invited it. If any man overstepped, it was about the worst thing he could do, and he was made to feel it."

There are still many vigorous young women—the kind Grandmother referred to as "tomboys"—who refuse to take a back seat for any man. They like to get out and work stock once in a while just for the fun of it. The neighbors of Bob Tisdale, who manages the Three T Ranch north of Casper, speak with approval and respect of his daughter Mike, who helps on roundups and takes her part with the best of the male workers. A girl like her is the exception in the cattle country, however. When ranch girls start to be young ladies, they take more and more to the corral fence.

The most extraordinary practicing cattlewoman in the country at this moment is Mrs. Elsa Smith of Idaho Falls, the only woman brand inspector in the United States. Mrs. Smith "grew up in the saddle." Her father came out from the Indian Territory in the seventies and settled in the Big Horn Basin of Wyoming. She was "just another cowboy," she says, until she was out of her teens. When she married John Smith and started riding herd on three sons and a daughter, her cowgirl days were apparently over, but these things are unpredictable. In 1937, John qualified as a cattle inspector for the Wyoming Stock Growers Association and Mother and the boys went right into the stockyards with him. Mrs. Smith worked mostly in the office, but she was still as good with a rope as with a filing cabinet, and made herself useful in the pens.

Time passed. The three sons went into the navy, and Mother worked harder than ever. The war ended; the boys came back; and that should have been her cue to retire to the kitchen. Instead she let Chief Inspector Russel Thorp know that she wanted to qualify as a brand inspector in her own right. Until his retirement in 1949, Thorp ran a regular school for apprentice inspectors, complete with quizzes and final examinations. For three years Mrs. Smith stood up to the curriculum, and finally "graduated" from a course in which only one out of five enrollees survive. By this time she was a grandmother.

At the June, 1949, meeting of the association she sat on the platform beside a son who had also qualified. She was dressed simply and modestly in regulation party dress and shoes—no cowboy flamboyance at all. As Russell Thorp presented her certificate, he called her "an exceptionally competent and capable inspector."

For contrast, look at the Hoag sisters of Uvalde, Texas. Mary, Lucy, and Daisy live on the old family place and run it with a little help from their brother who owns an adjoining ranch. Miss Mary, a fragile old lady in an ancient house dress, battered shoes with paint stains on them, a wide-brimmed blue cloth hat tied under her chin, and a pair of cloth gloves, hoes daintily at her pocket-size vegetable garden and recoils at the thought of doing cowboy work. "I have never ridden a horse," she declares. "Our brothers never wanted us to ride. They said it would spoil the cow ponies."

Human nature being what it is, one would expect that all ranch women would be dissatisfied some of the time, and some of them all of the time. They, too, have their crosses and frustrations. There must be many like Mrs. Lyle Henderson who lives, unreconciled, on a 700-acre place twenty-four miles from Watford City, North Dakota.

Lyle is a fine-looking, loose-jointed, self-possessed young man who is spoken of in the neighborhood as a hard worker and a fine cowboy. He met Haziel, a tall, handsome blonde, when he was a patient at William Beaumont Hospital in El Paso during the war. They were both Seventh Day Adventists and got acquainted at a young people's meeting one Sabbath. Haziel was fascinated when he told her about the 10,000-acre ranch in the rolling Dakota hills where his father lived—the church services they held in the schoolhouse—the cattle and horses roaming the pastures.

She married him. He bought the place adjoining his father's, put in a wonderful kitchen, selected bedroom and living-room furniture, planted the yard, and installed Haziel and their newly arrived daughter.

But Haziel is not too happy. She says she is lonesome so far away from everything. The mail road runs a mile or more from the house. The approach to her door is a trail through the fields, summer-fallowed now and butter-thick with wild mustard. The house is down in a hollow in the hills. Some people would call it cozy—but not Haziel. "If I could only see somebody's chimney smoking, it wouldn't be so bad," she says sadly.

She would like to return to El Paso and wants Lyle to pull up stakes and study engineering at Texas Western College—to get some training and become a worker in the Adventist church, or to do anything that will get her away.

Her mother-in-law comes over to see the baby—the first grandchild—and wishes she could do something to make Haziel happier. She thinks her husband may build another house for the young people across the road from their own—if the well diggers can get water. Meanwhile Haziel faces the hard adjustment that some women have to make to the conditions of ranch life, and those who love her can only wait, pray, and hope for the best.

83

How much unconfessed dissatisfaction is behind the steady townward movement of American ranchwomen there is no way of knowing. But more and more of them are shifting their base of operations to comfortable and sometimes luxurious houses in village or city. Some prefer town life; some go because the children have to have a place to live while they go to school. Either way, something like half the ranchwomen I know keep up some kind of quarters in town.

When Mother leaves the ranch for good, she settles into the routine of her churchgoing, bridge-playing, P.T.A.–organizing *comadres,* and loses her distinguishing characteristics. She may go back to the old place for a vacation or a visit. Her former neighbors may drop in after doing their buying on Saturday afternoon. She may still go to rodeos, conventions, and stock shows as of old. But the spell is broken. Her daughters marry navy lieutenants or automobile salesmen or manual-training teachers in the local high school, and Mother looks and acts like all the other mothers in the community.

Whether she lives in town or not, one cattleman's activity which the cattleman's wife will never give up as long as she can hobble is the business of attending stockmen's conventions. Local, state, and national meetings keep the men in fairly constant circulation, and if there is any way to manage it, Mother goes along. Mrs. George Green told me, a few months before her death, that she had missed only two meetings of Colorado stockmen in thirty years, and that in spite of none-too-robust health.

The ladies are just as serious about the speeches and discussions as their husbands are, and tend to grow restive when too many teas, receptions, and female blowouts keep them from the regular sessions. Furthermore, they are anxious to take an active part in whatever needs to be done. As a result the Cowbelles are ringing all over the range.

The cattlewomen of Arizona claim credit for the name. Mrs. Dean Bloomquist, present secretary of the local organization at Douglas, tells why.

For several years everybody had been getting together periodically at a hall on Mobley's ranch seven or eight miles from town. The usual entertainment was dancing; music and refreshments were provided by two or three families who pooled their resources for the evening. Finally it occurred to some of the women that an organization was needed for planning these affairs, and one of them suggested that such a group would be useful also in handling arrangements for the conventions and meetings that come to Douglas.

In October, 1939, sixteen women met and made the organization official. From then until now they have been known as Cowbelles. Mrs. Ralph Cowan, a large, blonde, fast-talking bundle of energy, was the first president. Mrs. John Cull, Mrs. John Murchison, Mrs. Frank Moore, Mrs. Will Glenn, and Mrs. Ella Glass are charter members who still labor for the organization.

Other states were not far behind Arizona in taking up the idea. The ranch wives of Wyoming were first to set up a state organization. Mrs. George Snodgrass wrote to the Douglas group to ask for pointers and to request permission to use the name. The Douglas women were not too happy about sharing their distinctive title, but didn't see what they could do about it. At least, they say, they have kept the one-word spelling while the other groups make two words of the name.

The Wyoming ladies announced firmly their twofold objective of developing their own "social well-being" and "promoting the welfare of the livestock business," and proceeded to run their membership to a total of five hundred. They give away money to worthy causes, stage an annual banquet while the men are having their stag party, and take charge of the memorial service for members of the association who have died since the last meeting. In 1948, Mrs. Snodgrass tried to make the ceremony a little more impressive by having someone sing *Beautiful Isle of Somewhere* while the names were read. As each one was called, a young lady placed a white carnation in a vase on the table. Mrs. Snodgrass isn't sure yet whether the men were impressed.

The Colorado Cow Belles were right on the heels of the girls from Wyoming. In 1941 the Colorado Stock Women's Association reorganized under the new name, and since then the idea has spread to local groups, the latest being the Plateau Valley

Cow Belles organized at Collbran. The hostess group at the annual meeting of the Colorado Cattlemen's Association at Grand Junction in July, 1949, was the Western Slope Cow Belle Club.

Their annual banquet was probably typical of Cow Belle gatherings all over the West. They ate baked ham, mashed potatoes, hearts of lettuce with thousand-island dressing, hot rolls, pineapple sherbet, and for dessert (Tee Simms reporting for the *Record-Stockman*) "a square of vanilla ice cream with the head of a steer in chocolate ice cream on top." A far cry from the beef, beans, bread, coffee, and dried peaches of the pioneer mothers.

During the meal George Currier sang "Wagon Wheels," Robert Currier played some Chopin, and Miss Eleanor Smith of Rifle contributed Wieniawski's "Mazurka" and "Ah, Sweet Mystery of Life." After dinner Preston Walker showed a color movie called "Rivers in Sand in Anasahzi Land."

"The dinner was attended by more than 170 of the best dressed, friendliest and most gracious women to be found anywhere," wrote Mrs. Simms. "The lovely outfits worn at this dinner were so numerous that there is not space to tell about them."

Mrs. Frank Bledsoe, retiring president, closed her remarks by reading the words of the Cow Belle song, part of which goes as follows:

> *Our home, our home on the range!*
> *In the spring the snow flurries soar.*
> *In July comes the rain and the flowers bloom again,*
> *Then in the fall, it is Heaven for sure.*

Idaho, Montana, Nebraska, Utah, California, and New Mexico have taken to the idea, and there are cowbelles now from one end of the range to the other. They are the most amiable and delightful ladies you could hope to meet—but there are those who look upon them as a menace. I have heard a prominent Arizona woman call them "potentially one of the most dangerous organizations in the country." She felt that as propagandists for the stockmen's views and desires they were far ahead of the men.

Mrs. Bloomquist reacts violently to this opinion: "When we organized, we didn't start out to battle any windmills. We wanted to have a good time, and you know the men aren't going to start any social activities. They stand around and talk about the weather and the cows." She admits, however, that the American National has asked the women to help enlighten the public on the cattleman's views, particularly questions involving the public lands.

This, of course, is a highly suspicious move. Many people are convinced that anything a cattleman wants is bound to be wicked and dangerous. If his wife backs him up, she must be wicked and dangerous, too.

As a matter of fact, America's ranch women will never be a peril to anything. They don't have time. Twenty-four hours a day are not enough to keep the woman's end of the ranch going and leave any leisure for subversive activities.

"T 7. Punchers

HE damn cowpunchers are getting pretty thin."
"It's mighty hard to get good cowhands. In another ten years there won't be any left."
"All the cowboys are truck drivers now."
"They're about gone. You can hardly get one to break a horse now. Not so long ago they couldn't wait to get at it."
"There isn't much need for them any more. So much land has been plowed up around here. When we need help, we call in the neighbors."

These discouraging opinions come from big and little cattlemen in Texas, Oklahoma, and Kansas. On the northern plains the story is the same—cowboys aren't what they used to be.

"When we branded early in July," writes Jennie Williams of Banner, Wyoming, "our crew consisted of our tenant, an ex-tree trimmer, an ex-dry farmer, a fourteen-year-old girl and one real cowhand. The only roping done was by our little seven-year-old Bobbie. A visiting marine and the hired man helped with the branding, and as the calves were cut into a corral away from their mothers, the 'wrastlers' simply walked up to them and flanked them. Please don't tell me that times haven't changed in the cow business."

We may regret the passing of the cowboy, but we need not be surprised at it. He belonged to a peculiar breed of men who were produced by the peculiar characteristics of the early-day cattle business. The industry has gone through revolutionary changes, and there just isn't any place for the old-time ranch hand. He would be about as much at home on a slicked-up modern ranch as Daniel Boone at a debutante party.

The old men who gather at the Stamford Cowboy Reunion, the annual meeting of former XIT hands, or Old Timers' celebrations in connection with Stampedes and Rodeo Weeks are under no illusions about themselves. They are well aware that they and their bunkies of fifty, sixty, or seventy years ago were not knights in shining armor—not ordinarily. They were hard-working country boys who did their work on horseback, endured the sweat and dust and thirst becuse they had to, took the boredom and loneliness and graybacks and granulated eyelids and "sugar lips" because these things were part of the only job they knew or cared about. They stayed single because there was no place for a woman in the life they led, even if they could get a woman, did without schooling and social life because there was so little of either available, and followed a rather grim frontier code of conduct because it was imposed upon them and because they lived longer and stood better with their fellows if they did.

Yes, there were younger sons of English earls who rode the range and wrote books about it in the afterglow of their careers. There were farm boys and dry-goods salesmen who got the bug, learned the hard way, and stayed with the herd. But cowboying on the frontier was a rugged business, and was practiced by rugged men. The voices of Hollywood and the pulp magazines to the contrary notwithstanding, the social status of the old-time cowboy was not high, his culture was deficient, and his manners were unrefined.

Texas cowboys in particular were thought of in other states, like Lord Byron, as "mad, bad, and dangerous to know." "The average run of these men," says a Montana writer, "was below par. Many of them left their native state because it was necessary for them to do so. They were expert cowmen, handy with their rope, light fingered in ranch and camp, exceedingly fond of card playing, a bit brutal to their horses, quiet at their work, but noisy and treacherous under the influence of liquor."[1]

These statements may cause me some trouble. For many of the historians of the range are sentimentalists on this subject and it is almost as risky to tell the truth in their presence as it is to talk about the deserters from the Southern armies at a meeting of the U. D. C. "Those ranks were composed largely of men with character and heart, of men whom future generations well may regard with pride," says Philip Ashton Rollins—a typical statement.[2]

Most of the confusion on this subject would be eliminated if people would remember that cowboys were human beings. There were all sorts and conditions. No two were alike. Put two or three of them alongside each other, and the point becomes clear and unmistakable.

At the Headquarters of the Warren family's South Ranch at Cheyenne you can meet John Ryder, a shaky old man with an expressionless face who sorts sacks and does odd jobs around the place. He is pensioned off now, but "he'd go crazy if he didn't have something to do." He was quite a fellow when they took him on forty years ago. In his youth he worked as a railroad dispatcher, but he sometimes took one too many and eventually got in bad with the management. Before he could be fired, however, he got wind of what was coming and took matters into his own hands. He sent out messages to everybody from the superintendent on down to come to the station for a special conference at 3:00 A. M. the next day. Then he boarded a train and headed west. He never heard what happened at the early-morning meeting, but it always amused him to think about the possibilities.

He was never in any trouble with the Warrens, but he sometimes got into scrapes when he went to town. Once he passed out after a lengthy celebration, and the boys took him in hand.

[1] John Clay, *My Life on the Range* (Chicago, 1924), 268.
[2] *The Cowboy*, 40–41.

They found an empty coffin, set it up in the back room of a grocery store, put John inside with a lily in his hand, and went off and left him. They felt repaid for their time and trouble when John came to, saw his situation, leaped from the coffin, and came forth waving the lily and roaring that they were trying to bury him alive.

Beside John Ryder we might place Billy Pardloe, "the Pitchfork Kid," who was a famous character on the Matador Ranch in Texas. He is remembered as a fine human being, reverent, honest, and much loved. An orphan from Kansas City, he did not even know his own name until he went back home in his manhood and looked it up. All his life he lived and worked on the Pitchfork and the Matador, and when he grew old, the managers took care of him as if he were a former president. That was what broke his heart. He was a fine roper and wanted to drag 'em in like everybody else. They wouldn't let him—they protected him. Other men as old as he were not so protected, and he never could understand why he was the one to be put at the easy jobs. Finally he died, still not understanding that it was because they loved and honored him.

Or take Jake Raines on the S. M. S. Jake was a short, fat cowboy with a full beard (if you can imagine it), who saved his money and became a minor capitalist. He lived as crudely as ever, never let a woman get near him, and kept on riding and roping even when he was comparatively rich. Eventually he died without even finding anybody in particular to leave his money to.

Ten thousand other examples might be cited—all different, and all interesting. And most of them would be dead.

The changes that have made crossbreeds out of the old cowboy strain have been coming on for a long time. The historians say they started with barbed wire, which eventually wiped out the open range and gave the traditional cowboy room to function only on the few remaining ranch "empires." The railroads which ruined the trail drives, the rise of a taste for baby beef, and the need for raising hay and row crops wherever possible— all these factors have been at work to take the cowboy off his horse.

The growing importance of the purebred industry is another strong influence. There is an almost infinite distance between the lean and leathery longhorn of the past and the animals which move with ponderous majesty through the pastures of today, geared in a sort of super-low. It is easy to run off twenty-five or fifty pounds of those expensive steaks and roasts by using rodeo tactics, and as a result more stock handling than one might think is done on foot.

William Compton, who has a small herd of good cattle on the old Liendo Plantation near Hempstead, Texas, says he hardly ever gets on a horse to handle them. He makes no unnecessary noises or quick motions, and his method works so well that they will come when he calls and even get into a cattle truck without argument. His Negro helper, who would give anything to let out a holler now and then, stands by and marvels. "Mister Will," he says, "You'se the best cattleman in this whole country, black *or* white!"

Then there is the invasion of the machine which has pushed the cowboy a little farther out of the picture. What corn pickers, combines, and manure spreaders are to the farm, bulldozers, cattle trucks, and squeezes are to the ranch. A couple of men can do today what a whole crew did in Grandfather's heyday—and do it without the perspiration. Hay loaders and stackers lighten the burdens of haying. Hot meals go out twenty miles to round-up crews who used to gather round the chuck wagon. There are stock tanks with kerosene burners to keep them from freezing and floats to keep them from running over. Two tractors dragging a cable can clear more brush off a pasture in a day than fifty axemen could in a month. Worth Evans is one of many who use butane gas to heat branding irons, eliminating the wood and smoke of the past.

Horace Hening says he thinks the pickup truck has done more to revolutionize ranching than any other one thing. The jeep, which climbs mountains like a goat and sneers at mud and snow, has gone a step farther toward changing the methods and the people. Along the highway outside Sulphur, Oklahoma, I watched a woman and a three-year-old child riding a jeep over cutbanks and ditches as they hazed a bunch of calves back into the barnyard. The little dogies were getting along faster than a horse could have moved them.

And the airplane has picked up where the jeep leaves off. Gloster Harral, out in the Pecos country, takes buyers up in his Stinson to look at stock, makes quick trips to Fort Stockton, seventy miles away, and even rides rustler patrol. When too many sheep began disappearing over the south fence, he let it be known that he would be flying that way almost every day, and the depredations stopped. Gus Glasscock of Premont helps the cowboys during roundups by spotting small bunches of cattle from the air. County Attorney Sam Smith at Meridian tells me that the Waco papers are actually delivered by plane to ranchers on the Valley Mills–Clifton–Meridian route. They get the *Times-Herald* twenty minutes after it comes off the press. The airplane is not just a luxury on the ranch any more. It is rapidly becoming indispensable.

The labor-saving devices which leave less work for the cowboy's hands to do are almost innumerable. Henry Koontz of Victoria keeps two radios in his cattle barn to gentle the stock. He says he is doing just what the cowboys used to do when they sang to quiet their herds. A South Texas rancher taught his cattle to come to his jeep when he sounded a siren, and was much pleased with the results until there was a bad automobile accident on the Falfurrias road. Two or three cars came together, strewing bodies up and down the highway. The police came from the north with sirens wide open, and the cattle charged up to the wire. An ambulance roared in from the south, screaming like a banshee. The cows went crazy, tore down the fence, and joined the party. Neighbors say that the inventive ranchman is now experimenting with other cattle-calling devices.

Here and there something that resembles the cowboy of yesterday can still be found. The big ranches in Texas, Oklahoma, Montana, and Wyoming which have not petered out or been subdivided are still strongholds of the riders and ropers—those who have not become professional rodeo performers. In the Panhandle of Texas and just below it—on the Matador, the JA, the 6666, the Three D—they still run the chuck wagon, sometimes almost all year round, and ranch just about as they did seventy-five years ago. This may sound more glamorous than it actually is. Modern-minded ranchmen think these big outfits are unprogressive and living in the bovine dark ages. "They haven't made any advancement since 1885" is the typical comment. Progressive or not, techniques which have almost disappeared in other parts of the country are still cultivated in such places and the cowboys are as skillful at range work as their grandfathers were.

There are still some good Mexican *vaqueros* in South Texas, a few first-class Negro cowboys along the Gulf Coast, and some handy Indian ropers on the Cheyenne River Reservation in South Dakota. But the all-round cowpunchers of the past are becoming victims of specialization. In their places you find tractor men, windmill men, herdsmen, and so on; and even the riders have a good many jobs to do besides riding. Milking, gardening, trucking, haymaking, cultivating, barn cleaning, and even egg gathering are taken in stride nowadays. Chores that the cowboys would once have scorned as degrading to his cavalier status he now performs without a quiver. As the farm has merged with the ranch, the cowboy has merged with the hired man.

He will never go all the way, however. There is too much tradition and pride behind the horseman of the plains, even though he rides only a few times a year. As long as he thinks he is a horseman, as long as he wears the uniform, as long as the rest of the world chooses to consider him somebody special, some flavor of the old gallantry will remain. He will never be just a workman. And when he condescends to tinker with the truck or stay up all night irrigating, he does it with a cheerful alacrity which seems to say, "I think there is some doubt about the propriety of my doing this, but I refuse to be little about it."

Something like an average "hand" would be Gene Harwell, a friendly young fellow in his twenties who works on the Flat Top Ranch at Walnut Springs, Texas. Flat Top is devoted to the breeding of prize-winning Herefords. It is big business, and even the minor employees on the place are in some degree specialists. Gene is in charge of the sales bull barn, where young bulls ready to step into adult bullhood are held, fed, and shown to prospective buyers. It is a job of some responsibility, for those young bulls have to look good, feel good, and stay in top condition to catch the eye of the pernickety breeder with a thousand uneasy dollars rattling in his pocket.

But Gene is not a cowboy. His father was a farmer near Walnut Springs, and although Gene handled stock from the time he was knee high, he has had little experience in range-cattle work. During the war he was in the navy, and after making a break with the country, he thought he might like to live in a big town. So he got a job in a machine shop in Dallas.

It didn't work. He didn't like the pressure of the factory. He didn't like the crush of people and the lack of freedom and privacy. When he had enough of it, he came back, asked Bill Roberts for a job, and was soon feeding the sales bulls.

"The job in Dallas paid three times a much," he confesses, "but I couldn't keep any of it. Here I have a good house, lights, water, chickens, a cow. The school bus runs past my door. I have a good car to get around in myself. My wife was a country girl and is used to the life—in fact, she likes it. We are better off here.

"I get up at four-thirty in the morning and put out feed. Then I come back to the house and get breakfast. The rest of the day I work around the barns till time to feed again in the late afternoon. I quit around six. There is no pressure on me. If I need to take time off, I can. The boss is my friend and lets me do the job my own way, though if the cattle aren't doing well, he soon notices. When there is an emergency—a cow that has to be treated or something—he expects me to go ahead and do it. It makes a man feel responsible."

Gene does not look a bit like a cowboy when he is at work, though he can wear boots with the best of them on special occasions. He wears heavy shoes, a cap, and a work shirt around the barns as he hauls in the feed, spreads straw, or works with the big panels that close the barn in stormy weather. The buyers and visitors who drive up he welcomes with friendly frankness. He talks neither up nor down to anybody, and is independent and democratic, as cowboys are supposed to be, whether they belong to the new model or the old.

The wages, working conditions, and manner of life of these

Feed Lot in Full Swing
J. D. Weatherbee fed out 36,000 head of cattle in this
lot near Tornillo, Texas, in 1948

ranch hands are a good deal different from what they used to be. On the one hand the men are much better off financially, but on the other their chances for independence and ownership are almost nonexistent. The base wage for ordinary work is somewhere around $100 a month, with living furnished. Earl Monahan says wages are higher in western Nebraska than elsewhere, starting at $125. He is one of the few who make a practice of varying the wage according to the ability of the worker. The two men on Emmett Horgan's ranch at Hermosa, South Dakota, get a flat rate of $125 a month. "Any scaling makes jealousies," Horgan says.

Locality makes some difference in wages. Henry Blackwell, of Cuero, Texas, has a Negro top hand who has been with him twenty-two years. The man gets a house, wood, a cow, and $22 a week.

At Liberal, Kansas, where the endless wheat fields are leased out for winter pasture, Louie Limmert employs a man and wife who draw $150 a month plus an interest in the crop.

Luke McOllough in the San Luis Valley of Colorado pays from $125, with a living, and this includes five acres of potatoes. One of his boys has a cow, but this is "not customary."

Laurence Fuller, who ranches near Wyola, Montana, pays $125 a month with everything furnished. He figures it costs $40 a month to feed a man now. His father-in-law, who runs 3,000 head of cattle, has a food bill of $1,000 a month.

Al Favour of Prescott, Arizona, says that a top hand in his country got $60 in 1940 and rates $150 now with all the usual extras.

Cowboys complain that employment at these figures is scarce. In September, 1949, Richard Aro told Tom Swearingen of the *Record-Stockman* that "jobs were hard to find and that rates of pay had shown an alarming decline." He had found only one riding job, and it was snapped up by somebody else. Next best was a hay-stacking offer near Fairplay at $75 a month.

Even $75 a month seems a far cry from the day in 1900 when Jim McMurtrie went to work for the J. A. outfit at $25 and grub. But the new-model puncher is not as far ahead as one might think, for all he has is his wages. Jim McMurtrie is now a wealthy retired cattleman nursing his arthritis in a fine house at Clarendon, Texas. How did he do it? Well, like all ambitious punchers he wanted a brand and a ranch of his own, and little by little he got them.

The old-time ranchman was no dog in the manger. The Reverend J. Stuart Pearce, now running a few cows near Catarina, Texas, worked for John Blocker in the days when cattle kings liked to see their men get ahead. He says that Blocker never paid his men entirely in cash if he could help it. He gave them a few heifers, and when they had a hundred head, he would lease them some land to give them a start.[3]

Now, declares J. Y. Crum of Weatherford, "most of them wouldn't let a man have as much as a chicken of his own. There won't be a horse or cow on the place that doesn't belong to the owner. When a man walks off, he leaves everything. Tom Saunders lets Walker Good run twelve head, but it is the only case like that around here."

With high-priced land and cattle, little capital, no free range, no mavericks, and none of the free-and-easy methods of the past to give him an advantage, the puncher has about as much chance to become a cattleman as a rustler has of getting to Heaven.

Without so much spur to his ambition, however, the cowboy of today has some compensations that his predecessors did not have. He can get into town oftener, if he wants to, and most boys like to feel a sidewalk under their boots come Saturday night. The old days when men on the big ranches did not get off the place for months are long gone. Their pleasures in town are what you would expect of ordinary Americans—movies, soda fountains, and church suppers for some; movies, barrooms, and pool halls for others; movies, home cooking, and the companionable interest of a wife or mother for the rest.

For the cowboy is no longer a runaway youth who thinks he might go back home "when the work's all done this fall"—a lone wolf who lives exclusively with men and is tongue-tied in the presence of "good" women. Cowboys get married now, and the

---

[3] J. Frank Dobie is skeptical about such generosity. "The average cowman did not encourage cowless cowboys to 'start a brand.'" (*The Longhorns* [Boston, Little, Brown and Company, 1941], 53.)

ranch owner has to make provision for doubles. Some owners still prefer single men, but many more have learned to tolerate, and even prefer, the married hand. "If I were starting over," remarks Clyde Buffington, "I would begin by building a row of tourist cottages on the ranch. They all get married now."

This does not agree with the prevailing theory. Magazine articles, written by men who specialize in telling people what they want to hear, are forever assuring us that the cowboy "hasn't changed much. He is lean and loose limbed. He lives out in the open, works sun-up to sun-down, rolls cigarets, wears boots every waking minute, and never walks more than a few yards if a horse is handy." Cowboys "practically never have days off and most of them never get married."[4]

This is pure folklore. Johnny Stevens, present manager of the Matador, says his men average better than thirty years of age, and most of them are married. He prefers to hire older men, even for wrangling horses, because they are more apt to know the country well and can be depended on. Just recently one of the hands got lost in those vast pastures and a crew had to be sent to look for him. Greenhorns are more trouble than they are worth. Naturally most of the employees are local boys, and they habitually marry and settle down in the little town of Matador.

The hands are always welcome to eat at the ranch, either at the cookhouse or the chuck wagon. But even when the wagon is thirty miles out from town, some of the men prefer to ride in every night for the sake of spending a little time at home.

Where the old-fashioned, unmarried cow hand survives, he lives in the traditional way. His home is the bunkhouse, which may be anything from an old shack heated with a wood stove and no toilet facilities to a modern structure with all conveniences. But here again some old notions need to be brought up to date.

The cowboy of history, says Edward Everett Dale, "was in most cases an excellent housekeeper. . . . Only a bum, it was asserted, was willing to live in the midst of dirt and disorder."[5] This may have been true fifty years ago. It is not true now. In

---

[4] "Texas Ranch," *Life*, October 21, 1940, 75.
[5] "Cowboy Cookery," *The Hereford Journal*, January 1, 1946.

general, bunkhouses are not places that the ranch owner is particularly glad to show off. The boys live in womanless disorder and don't bother much with picking up and keeping things neat. And even when special efforts are put forth to provide them with a decent place to live, they don't always appreciate it.

Probably the most grandiose effort to civilize the cowpuncher by putting him in proper surroundings was made some years ago on the Pitchfork Ranch (three hundred sections, six line camps) near Guthrie, Texas. It belongs to Eugene and Gates Williams of St. Louis, famous for making Buster Brown shoes, who come down in the quail season for fifteen or twenty days and let God and Dee Burns take care of it the rest of the time.

Emma Williams, a sister, decided that the ranch should put up a bunkhouse to outshine all others. The manager was skeptical—said the boys wouldn't take care of it—but had to give in. "All right," he growled. "I'll build 'em their damn bunkhouse."

The result was a two-story frame building painted a rebellious blue-gray and looking like an Odd Fellows' hall in the lumber district of Michigan. Inside, the walls are paneled in natural wood, which does not look so natural now. The halls have not been swept for a long time. On the second floor landing there is a sign which says "Trash Downstairs." Underneath is a trash can which was long ago filled to overflowing and is now disgorging copiously on the floor.

The rooms are fairly large with beds and built-in chests of drawers for two boys in each. The beds are miscellaneous—army cots, iron double monstrosities from the early nineteen hundreds, and so on. Apparently the boys furnish their own bedding, and have provided everything from G. I. khaki quilts to plaid cotton blankets. No spreads—and most of the beds unmade. The occupants leave their boots and spurs on the floor and their garments on the furniture. The closets have some good clothes in them, every hat a Stetson and the boots of fine quality. A variety of pin-ups adorn the walls, but in only one room are the decorations prevailingly nude. Family photographs and pictures of girl friends are the rule.

In one room a boy with glasses and a thoughtful face is stretched out on the bed with his boots on listening to a radio.

"There's a radio in every room," Assistant Manager Coy Drennan tells me.

"Do you ever hear any of the boys sing?" I ask the prostrate puncher.

He looks astonished at the question. "Cowboys don't sing," he replies positively.

"I think some of them carry radios on the saddle," Drennan adds.

Downstairs the cistern has leaked into the basement and flooded the heating plant. The bathtub in the shower room needs a cleaning worse than West Texas needs rain. There is a community room with a heating stove and some bunged-up furniture where the boys are supposed to gather for companionship, but the chances are they gather in some more cheerful place.

One understands why Emma Williams gave up, and why the bunkhouse was never completely finished.

The foreman who built the place quit, partly, they say, because of unpleasantness over this white elephant. He took his assistant, Johnny Stevens, with him, and Johnny went on to become manager of the Matador, but he remembers that bunkhouse on the Pitchforks. He kept it policed up when he was in charge, he says. The boys took turns cleaning it. There was a list of duties assigned to particular men, and if anybody was slack, they brought him before a kangaroo court. "We cracked their tails with a board if we found them guilty," Johnny recalls. "I can't see any excuse for having things filthy."

Archie Sanford's bunkhouse on the ranch at Alcova, Wyoming, is an old log building put up in very early days. It is occupied by one man, and he leaves it a mess. The floor is invisible under boots, papers, horse furniture, and abandoned clothing. The brightest spot in the room is a sign over the radio on the table: "Jesus Saves." Otherwise the room is hopeless.

When the boss is particular, or when a woman is given the job of sweeping up, you find a neat bunkhouse. But don't expect it under ordinary circumstances.

The cookhouse is another matter. In the big room where from two to fifty men gather for meals, everything is shipshape. On a big ranch all the conveniences from a meat slicer to a walk-in icebox promote efficiency. Owners and managers are proud of the way they feed, and those who have given the matter any thought will tell you that good food and inferior wages are more acceptable to the men than inferior food and high wages.

On the Pitchforks the bell rings at a quarter to six in the morning and breakfast is on the table at six. Supper is due at six in the evening, and the men were drifting in from the pastures and barns at five the afternoon that I was there. Twenty plates were turned upside down on the long table in the dining room, which used to be the living room of the old ranch house.

A good story is told about the Pitchforks which shows how the culinary department is regarded by the hands. The cook in days not too long gone was a motherly woman known as "Miss Sallie." She took care of the boys for years—mended, patched, comforted, and cared for them in all possible ways. When she finally became almost too old to work, Mr. Molesworth, the foreman, hired a Fort Worth boy to help her. He turned out well. Miss Sallie liked him and he was a good worker, but in about a week he came in and asked for his time.

"What do you want to quit for?" Molesworth asked him. "I thought you was doing fine."

"I just want to quit."

"All right, but I want to know why."

"Well, I'll tell you. I go down to milk and come back by the bunkhouse with the buckets, and the men call me 'Miss Sallie.' I don't like it and I'm going to quit."

He did, too.

Molesworth took him to town, and when he got back he called the boys in for a little chat. "Boys," he said, "you all know what Miss Sallie has done for you. She's cooked good meals—not just beef and beans but pies and things. She's tended every one of you when you got sick or hurt. It's too bad if you can't help her out now in her old age. And now I'll tell you what I'm going to do. The first man that calls the new boy 'Miss Sallie' is fired no matter who he is, old or new. Understand?

They understood. So a new boy came along and lasted another week. When he asked for his time, Molesworth was ready to fight a buzzsaw.

"Son," he asked, with a wicked gleam in his eye, "has anybody called you 'Miss Sallie'?"

"No sir," said the boy.

"Then what's the matter?"

"Well, I go down to the pen and milk, and every time I come back by the bunkhouse with the buckets, they all get up and take off their hats. I'm damn tired of it and I'm going to quit."[6]

Thanks to the Miss Sallies of the cattle country, and to the gadgets they have learned to operate, the modern cowboy is better off than his grandfather in the matter of food. Where the chuck wagon still runs, the traditional camp food is provided, with beef, beans, potatoes, biscuits, dried fruit, coffee, and sirup as staples. But canned foods have added variety, and the ranch garden helps out. No longer can the cowboy say, as he once said, "I have a thousand things to eat tonight—beans!"

It would be almost 100 per cent correct to state that the cowboy eats like everyone else now—or maybe a little better. He has oatmeal and orange juice for breakfast once in a while, bacon and eggs frequently. Steak for breakfast is not as common as it was before the price of meat went up. Green salads, pickles, preserves, and cookbook desserts appear on ranchmen's tables from Brownsville to Billings, and the hands get their share.

Not every cowman gives up to a citified diet without a struggle, however. Food can be a little fancy—yes! But it had better not be too fancy. If it gets too far down that road, the old cowboy starts making up names for things. George Snodgrass annoys his wife by asking her to pass the train wreck when she mingles fruit with red jello, and he raises her dander higher by calling corn flakes "skillet scabs." To such men rice pudding is "moonshine," and the meringue on a lemon pie is "calf slobber." Probably the ultimate in verbal rejection was achieved by the cowpoke who reacted to a plate of jello in these words: "I'd as soon ride into a west wind with a funnel in my mouth."

Something might be said also on the subject of coffee. Elbert Hubbard defined a ranchman as a person who had ten thousand cows and put condensed milk in his coffee. This is more or less

[6] John Hendrix tells this story briefly in *The Cattleman* for August, 1942. Harold Bugbee of Clarendon contributed the version used here.

accurate. But there is still a strong prejudice in the minds of some ranch owners against putting anything in coffee at all, except maybe a little sugar. Buck Pyle of El Paso, who manages the vast West-Pyle holdings in Texas and New Mexico, loses his composure over it. "The only thing I can't stand," he says bitterly, "is a lot of these damn condensed-milk cans sitting around at the chuck wagon. They can have all the coffee they want, with sugar in it or not, but I won't buy any goddamed canned milk for them." And Buck is not by himself in that respect.

Cowboys are not all on the same level today, any more than they were in the trail-driving epoch. A man can still work up to the position of top hand, straw boss, or foreman. These terms are elastic and mean different things on different ranches, but it has been said with some truth that the pattern for leadership among cow hands was set by the Mexican *caporal.* Among the Mexican cattlemen who pioneered the business long ago the social system was still feudal, with *ricos* at the top and *pobres* at the bottom, and nothing much in between. The *caporal* belonged with the people at the bottom and never felt that his position brought him into anything like equality with his *patrón.* Knowing his place, he kept it with dignity. In South Texas he still exists.

I spent a night with Rafael García on Alvino Canales' ranch a few miles from Premont, Texas, in order to find out about these things. Don Alvino has four or five ranches, including the small one of 2,400 acres where Rafael lives—all of them under Rafael's supervision. His job is complicated by oil wells and oil workers at every turn of the trail through the mesquite pastures, but they do not change the Mexican pattern of his life.

Alvino's son Gus drove me out to the headquarters—a neat white house deep in the high brush. There was a windmill on one side, a Quonset garage on the other, and a sprinkling of white leghorn chickens in between. Rafael's wife, Sara, a young woman from Mexico City, met us at the door, obviously under some strain at a visit from the son of the *patrón* and an American visitor. She seated us in a small living room with studio couch and matching chair, coffee table, and wedding picture of herself and

Rafael on the wall—Rafael looking slender, intense, earnest, and a good deal older than she, but *muy hombre* in his black wedding suit with a spray of lily of the valley in his buttonhole.

After polite inquiries about the health of various members of Gus's family, Sara served us coffee and sat primly in her chair, hands folded in her lap, speaking only when asked a question.

Don Alvino keeps a room in the house where he takes a siesta in the middle of the day when he is on the place. That was to be my room. By the time I got my bag in, supper was ready. I washed at the kitchen sink, threw the water outside, and turned around to meet Sara's uncle Pedro Vela, a comical middle-aged Mexican bachelor with a round face and a twinkly eye, and her twelve-year-old brother, Juanito. As we ate the beans, sweet potatoes, and corn bread which Sara had for supper, I found that Juanito was very eager to be an American and had an *English for Foreigners* book under his elbow. We went to work on it when the dishes were cleared away, with special attention to "this" (not "theese") and "am" (not "em"). Sara hovered around trying to learn something, but did not join in the conversation as an American woman would.

At bedtime they showed me the outdoor *escusado* and left me the family flashlight. In the morning we had eggs and beans after an hour's hunt for the eggs. There was one apiece for everybody but me. I had two.

When I left, Rafael refused to take anything for the trouble I had been to him and his wife. Señor Canales provided everything. I was his guest.

Later I learned that Rafael had been an independent operator himself. Once he had owned four hundred acres of cotton land in Mexico and was worth as much as 100,000 pesos. Then the *Agraristas* moved in, and Rafael fought them. Furthermore, he was a good shot, and as a result he had to give up everything and cross over into the United States. Fortunately he was born on the American side and could legally demand sanctuary, but he had no chance of starting in again for himself. Now he holds a position of trust in the powerful Canales organization, and that is no small distinction in its own way. But he does not slap Don Alvino on the back and offer to buy him a drink. There is great kindness between the two of them, but not the kindness of equals, all according to the ancient customs of Mexico.

The modern American counterpart of the *caporal* could be any one of the four Bond brothers who work on Miller Ainsworth's 10,000-acre ranch near Luling, Texas. General Ainsworth's grandfather took over an old Spanish grant in 1836, but there are no Mexicans on it now. The hands are all Negroes except the Bonds, who do the straw bossing. They are lean, hatchet-faced men who never step out of their roles. When they have to dress up, they put on clean levis and denim jackets to go with their neat boots and flat-crowned hats. They have dignity but no airs. Occasionally the Ainsworths take some of their friends out to the big tank in the south pasture for a fish fry. The Bonds know they are welcome to join the party, but they keep a little bit to themselves. You might say that they join the party when the party joins them, especially if there are women present. When the men drift over and gather round the mesquite-wood fire which the Bonds have built, they talk freely and enjoy the social occasion in a quiet, unaggressive way. Obviously somewhere in a middle station between the owner and the hands, they stay on their own side of the fence, and their manner discourages any undue familiarity.

Another cowman who stands apart from the ordinary hand is the herdsman, a figure unknown in the early days, who has come in with the expansion of the purebred industry. He is not usually educated, but all the practical know-how that anybody can have about cows is compressed into a little lump in his shaggy skull, and consequently he receives much respect. The problems of feeding, culling, and showing fine herds are in his hands. He travels to all the stock shows, knows all the big breeders in the country, and is full of behind-the-scenes stories about everybody who is anybody in the cow business. Some of the best herdsmen have been Englishmen or Scots, maybe because people in Europe are willing to work harder at mastering their skills.

On the Turner ranch near Sulphur, Oklahoma, an old Herefordshire boy named John Blenkin is a sort of herdsman emeritus ("consultant," they call him). A small, comradely man with a serious, elfin face that suddenly becomes delightful when he smiles, he sits in the office or naps on the leather sofa, goes out occasionally to look at the stock when his advice is needed, and talks gently about his memories of a vanished England. He remembers the bicycle trips he used to take to see his uncle at Cardiff—the beautiful green English countryside which belonged to the lofty people in the big houses—the game killing which was about the only way poor people could express their defiance of social distinctions.

He chuckles over a friend of his who had picked up a couple of hares on a big estate and was confronted by the owner on horseback.

"My good man, what have you there?"

"Hares, sir."

"Do you know that this is my land, and that you are poaching?"

"Well, maybe I am."

"I'll tell you what I'm going to do. I'm going to have you up before the court for this. You have no right to come on my property."

The poacher flared up: "How did you get the land, anyway, I'd like to know."

"My ancestors fought for it."

"Well, then, get down off that horse and give me a chance at it."

When his learning (and poaching) days were over, Blenkin came to America for a visit—just to see the country. But he never went back to live. His second job was with the Hazford ranch, where he stayed for eighteen years. He was in charge of all the breeding cattle when the place was sold to Roy Turner, now governor of Oklahoma. Turner acquired the best part of the Hazford stock as a foundation for his own purebred herd, and he brought Blenkin along to Oklahoma as part of the bargain.

Outside the ordinary run of punchers, also, is the ranch cook. Humble but useful, he makes quite a difference in the success or failure of a ranch enterprise, as many big operators are finding out since cooks began getting scarce. The Swensons at Stamford, Texas, have stopped running a chuck wagon, mostly because they cannot find a good camp cook. The Matadors keep the chuck wagon going, partly because they have one. On many ranches women have taken over the cook's job, and in emergencies one of the regular hands may step in if he has any idea what to do. On the Pitchforks a man and his wife manage the kitchen, but when I was on the place a young cowboy who weighed three hundred pounds and was getting too heavy to ride was temporarily in charge.

There is a real ranch cook at the Lamb's Head Ranch near Albany, Texas—a historic spread now operated by Princeton-bred Wat Mathews. Jake is middle-aged, self-confident, and a velvety black in color. He works in a fine new building ("We've been working on it forever," says Wat) which is a combined kitchen, dining room, and clubhouse for the men. It is about thirty feet long and twenty feet wide with a varnished-wood ceiling and lumalite fixtures. In the northwest corner Jake has his place. He is making biscuits, laying them in a well-greased pan with a black hand which has most of the thumb missing. He is getting ready to put them into a big wood stove against the north wall. Against the west wall is a water heater and an electric range.

I ask him why he doesn't use the electric stove.

"The sonofabitch ain't got no room," he says, opening up the oven to demonstrate. Then he flings wide the door of the wood stove, revealing a yawning cavern that would hold a beef.

"I gotta get five, six pans into the goddam thing if I'm gonna feed fifty people."

"What do you give the boys?"

"Ever'thing. Potatoes, corn, beans, turnips, jelly."

"How about meat?"

"Oh yes, we give 'em meat . . . steaks, roasts. . . ."

"For breakfast?"

"Naw, I give 'em bacon for breakfast—hog jowl. That's a pretty good kind of bacon."

"Too fat, isn't it?"

"Naw, it's got lean in it."

Jake has been with the outfit eleven years, speaks and cooks as he pleases, as ranch cooks have always done.

In one particular way the cowboy has not altered. He still

loves to exchange experiences, spin yarns, tell tales, and "swap lies." He likes best to recount what has happened to him or to people he knows, with or without exaggeration. The windies that used to be heard around the campfire or in the bunkhouse may not be as common as they were, but people who work with cows, from the millionaire owner to the humblest roustabout, are still full of anecdote. Sometimes the tale is sad; sometimes it is smutty; sometimes it is funny. Whatever the mood, somebody is telling one all the time in the hotels, clubs, barbershops, and cafés where cattlemen hold forth and do business. Horses, dust storms, blizzards, rattlesnakes, bucking horses, the peculiarities of women, the local politicians—anything that has happened or might happen is good for a story.

Somebody mentions the possibility of a snake crawling into a man's bedroll, which is enough to start "Scandalous John" Selmon. "No, I never had one crawl in my bed," he says. "But I slept one to death one time.

"It come up a little rain about bed time and I was in a hurry to get my hot roll fixed. I had an awful good tarp that I wanted to get under, so I jerked the rope loose and spread her out fast. I went down on my knees right on top of it to get everything arranged, and as soon as I could I got inside and never turned over till morning. About daylight I rolled out and started to tie up my bed. We always rolled 'em up from the head, and about the time I got it half rolled up I see that snake coiled up and mashed flat as a flapjack. God, I hollered! I was right at him. Everybody in camp come running to see what was goin' on. I guess he crawled up to my roll to get out of the rain, and never had no chance to move when I unrolled my tarp and soogans so fast. And then I went down on him with my knees. He was dead all right. But I bet I wouldn't have slept a wink if I had known what I was lyin' on. He sure scared the fool out of me!"

There is always a run of stories about the old-timers, dead or alive, who have lived in the community. Let the name of Clay McGonigle come up at Clarendon, Texas, and Pinky Price is reminded of the time Clay's daddy had trouble with a neighbor over a fence. They sparred for months and never got anywhere. Finally old man McGonigle reached the breaking point. "Goddam you," he choked, "I'm seventy-four years old, but you just step outside with me, and if you're man enough to whip me you can get your picture taken and sell all you want back where I come from at four dollars apiece!"

The picturesque metaphors still pop out naturally. "I was ridin' the south pasture fence when I jump this old bull. The brush was so thick a lizard would have to run three hundred yards down the road before he could find a place to turn off. The bull had been hidin' in there a long time, and he was sure goosey, too. I figured I was going to be as busy as the back end of a shooting gallery if I got him out—probably split my pants doing it—but a tubful of water and eight hundred cows just don't go together. He had to be got out of there, so I got my rope ready and taken after him. I'd sure as hell make a spoon or spoil a horn."

Those phrases all came out of the mouths of living cowboys, though not so close together.

Once in a while an old-fashioned cowboy "windy" creeps into the conversation. I heard one in the bar of the Elks' Club in Del Río from former Sheriff Red Hawkins, a straight-faced humorist with the nose of a tapir, the eye of a moneylender, a set of snaggle teeth, and a belly that sagged out over his belt buckle. He leaned back against the bar and told about the stranger he came across one time during his ranching days.

"Are you shipping any cattle now?" this fellow asked him.

"Why yes, some. But I'm having a hard time because my steers have got too fat," replied Hawkins, pulling hard on the man's leg. "They've got so big I got to have a separate car for each one."

"M-m-m. Those are certainly big steers. I don't think I ever saw anything as big as that in my life."

"No, I bet you didn't."

"I *did* see something big once though."

"Yeah?"

"Yes. It was a hotel. It was so big the dining room seated 25,-000 people. Never saw such a dining room. Almost unbelievable. The place was so big they had railroad trains running around on tracks to serve the guests. One special train had seven cars for nothing but pepper sauce and catchup.

"Half a dozen niggers were working on that pepper-sauce train, and one day one of 'em fell off. They never could find him. But a few days later they noticed a black object in the water gauge on the engine. He'd fallen into the water tank and come up that way.

"Out in the back yard was a sawmill they used for nothing but to cut up toothpicks for the guests. And they had six or seven more niggers with cowhides strapped to their feet dancing around on the stove to grease the griddles for flapjacks. Those niggers on the stove . . ."

"Just a minute," said Red. "You needn't go on. I don't own a ranch and I haven't got even one head of livestock."

I 8. Drifting Cowboy

I DIDN'T NOTICE Smoky Downing at first. I was in H. P. Pollard's saddle shop just off the main street of Douglas, Wyoming, and peppery old Mr. Pollard was telling me about his business.

"I've been here thirty-two years, and there used to be a hundred cowhands that bought from me," he was saying, puffing a disgusted breath out through his clipped white mustache. "But there aren't many left now. And they're all jumpy about the price. I don't hardly carry any Navajos any more. I used to get eight to twelve dollars for them. Now I have to charge sixteen to twenty-four, for the same stuff. They back off and say that's too much.

"Boots? Well, I used to buy a lot of Hyer's boots, but they ain't what they was before the old man died."

"Western makes a damn good boot," said a voice behind us.

Then I noticed him. A middle-sized, slender-waisted cowpuncher in blue denim pants and jacket, an old broad-brimmed straw hat on his head, worn black boots on his feet. His face was rugged but somehow shy, and was a rich brick red under a three-day growth of sandy stubble. His teeth hadn't been brushed in a long time, and he needed a hair cut—bad. He was smoking a tailor-made cigarette.

109

"Did you want something?" Mr. Pollard inquired without eagerness, well knowing what the answer would be.

Smoky grinned disarmingly. "I'm waiting for the county agent," he explained. "He's got a job for me on a ranch somewhere, but he won't be back for a few minutes so I come in here to smell a little leather." He picked up where he left off.

"I've tried 'em all—these boots. When I want a new pair, I look around in a store and if I don't find what I want, the storekeeper will send my measurements in to the factory. Sometimes I draw a picture of my foot, measure my instep, heel and ankle, and send it in myself."

"That's what we call a custom boot," said Mr. Pollard. "There aren't many of the old-time bootmakers left—the ones that make them all by hand. Most of the boys get their boots custom made now. They cost from thirty-two fifty up."

"These," Smoky remarked, holding up the right one and pointing to the vamp from which the leather was peeling, "cost me twenty-eight dollars and they ain't worth a damn. If that had been a custom-made boot, it would have gone right back in their face."

"They'd have made it good, too," Mr. Pollard told him. "They always make good when a boot isn't right."

"Nothing is as good as it used to be"—Smoky included the whole of our mass-produced civilization in a sweeping gesture. "Take levis. I won't wear 'em any more. They shrink till you can't wear 'em. I had a friend that used to buy them two sizes too big. He was bigger than I was, and when he washed 'em he used to make me wear 'em to stretch them out so he could even begin to get into 'em. I wear Lees myself. They don't shrink quite so bad, but they do fade. My underwear used to be blue all the time, where they faded off, and now my hands get blue because they fade on the outside too. And they don't last. Levis used to wear six months. Now thirty-forty days is about as long as they hang together—not long enough to pay you to wash them. I wear 'em and throw 'em away."

"They don't make chaps so good either," Mr. Pollard said. Everybody that works in leather makes 'em, but a lot of them ain't much good."

Cattle-country Cafeteria
Hands in front of the chuck-wagon tent

Photograph by Bill Giles

"I've got a pair of yellow chaps I wouldn't take any money for. I wear 'em a little short and let 'em hang down from the waist so I don't carry a lot of leather. I was working up in Montana when we took on a new fellow and the boss told him to go out in the brush one morning and bring in two-three cows. He didn't have no chaps or nothing. I was doing open work that day, so I lent him my yellow chaps. When I come back that night, he was gone and the chaps was, too. He quit at noon, rolled up my chaps in his bedroll, and taken them with him.

"It took me two years to get them back. I was walking down the street in Cheyenne one day in the fall when I seen him standing on a corner. He looked just the same, but I looked different. Last time he saw me, I had a beard and work clothes on. Now I was shaved and was wearing gray sport pants. I eased up to him and says, 'Howdy.'

" 'Howdy,' he says. 'Howya doin'?'

" 'Fine,' I says.

" 'Know anybody that wants a puncher?'

" 'I sure do,' I says. I did, too—I wasn't foolin' him. 'What kind of outfit you got—got a saddle?'

" 'No,' he says, 'I ain't got a saddle. I've got my spurs and a bed.'

" 'How about chaps?'

" 'I've got some chaps,' he says—and that was what I wanted to hear. Pretty soon he says, 'Come on up to my room and look at my outfit.'

"So I went.

"Sure enough, there were the chaps. But they didn't look exactly like mine. They were brown and had some ornaments hung on 'em. And he had laced a couple of pockets on with a regular shoelace. Me, I never have anything on a pair of chaps that'll catch in the brush.

"I held them chaps up in front of me, and they looked like my size. Hell, they ought to of—they was mine.

"I had my knife in my hand and I cut a seam on the side. There was yellow leather underneath. He had *painted* them chaps with ordinary paint.

111

" 'Brother,' I said, 'you don't know me, but I know you. Do you remember two years ago you run out of a cow camp and stole these chaps?'

" 'I never stole nothing in my life,' he says.

"That made me mad. I still had my knife in my hand and I showed it to him. 'You sonofabitch,' I said, 'you stole them chaps from a man that done you a favor, and I ought to cut your throat.'

"Well, he owned up to everything and I walked out with the chaps. But what did he want to paint them for? He could of used leather dye if he wanted to change the color."

"Maybe he thought the paint would turn water and keep out the rain longer," surmised Mr. Pollard.

"Well, I'm still looking for something to take the paint off of 'em."

It didn't take much more to get a picture of Smoky's life history. Havre, Montana, is his home. When he was just getting into his teens, the family was on relief. At fourteen he went to work for a man who was running cattle and making a little extra by rounding up and breaking wild horses. He paid Smoky five dollars a month and his keep to wrangle horses for the regular hands. Smoky did his job on foot, and sometimes bareback on a horse.

He asked the boss repeatedly if he couldn't ride for him. The boss would say, "Sure, when you get a horse."

It was a pretty big problem, but Smoky saved his money and finally had a few dollars. He approached the boss and said, "I got a little cash, and I want to buy a horse."

"Well, you can have jughead, there," the boss said.

"How much do you want for him?"

"I want a dollar for every year he's old."

"How old is he?"

"Fifteen. Next year he'll cost you sixteen dollars."

"You can keep him," Smoky said. But he reflected that Jughead had once been a good horse and was not all gone yet. Finally he bought him, patched up an old saddle that a cowboy friend gave him, and went to riding.

His salary immediately jumped to twenty dollars a month.

And like the rest of the hands he got one of the wild horses out of every band they caught.

That was how he met the love of his life. He noticed him one day in one of those little knots of wild-eyed, terrified creatures—a shiny black with four white feet, a white face, and two wall eyes. Smoky loved him at once. The more he looked, the more he loved. Finally, like any lover, he went to pop the question.

"I want to buy another horse," he said to the boss.

"Take your pick for twenty dollars," the man told him.

Smoky had only fifteen. "Tell you what I'll do. I'll give you fifteen now and the other five next Saturday."

"All right, which one do you want."

"I'll take the black over there."

"Him?" The boss could hardly believe his ears. "You can sure have him. I'll *give* him to you."

That sounded bad, but it was worse when he got home with his horse.

"My father was about ready to kill me," Smoky chuckles. "He was so mad he hollered at me, 'You'll shoot that horse before you're through! You ought to know that those four white feet mean weak hoofs. A white-faced horse is never any good. And with those wall eyes he'll be mean as hell. I'll bet you he'll strike and kick and bite right now.'

"I knew he would strike and kick," Smoky says, "because I'd seen him do it, and I was pretty sure he would bite, too. So I didn't say anything. But I started in to gentle him. I worked with him two months before I swung on his back. And he never pitched a bit. He didn't pitch till after two years."

"What started him then?"

"Stallion. When they don't get exercise they get high spirited and pitch a lot. That was how come I lost him."

Smoky stops for a minute and great sadness comes over his seamed red face. Mr. Pollard and I wait respectfully, for this is tragedy coming up—we can feel it coming.

"Well, he pitched so much I decided I wanted a gelding," he goes on, hurrying a little to get the painful part over with. "So one day I cut him. Next morning I went to the corral to see about him and there he was dead. Died right in the same corner where

113

I left him. I dug a grave right beside him and buried him six feet under.

"After that I was through there. I got together everything I had, including my string of horses, sold it all and went to town. I stayed drunk for six months.

"That was the best horse I ever had or ever will have . . . .

"After that I went out to Nevada and worked for a wild-horse outfit out there. I just drifted out and drifted back. I guess I'm restless. I like to be on the move."

"Don't you ever get tired of it and want to settle down?"

"By God, sometimes I've been so sick of it I thought I couldn't stand it another day. But after I rested up a little and sort of got back on my feet, I wanted to be on the move again. I've been everywhere—Montana, Canada, Nevada. Four years is the longest I ever worked in one place."

"How about getting married?"

"No sir, I'm not *about* to get married. I'm married to my horses."

It was time for us both to go. Smoky was due in the county agent's office. The last time I saw him he had got his bedroll and war bag out of his hotel room and was carrying them on his shoulders, towering high above his head. He looked a little strange and out of place among the filling stations and appliance stores and plate glass. People turned to stare at him.

He didn't care. His honest, unshaven face was set straight ahead. Tomorrow he would be on a new range.

I 9. Hitchhiker

I PICKED UP Charley Cocanower about fifteen miles north of Seymour at ten-thirty in the morning. He was standing beside the road away off there in the middle of the endless brush pastures of North Texas, thumbing vigorously at every car that went by. It seemed to be a sort of crusade or campaign that he was waging singlehanded against the indifference and callousness of Humanity. You gathered that he didn't have much hope of success, but was going to fight it through to the end. You could see him from half a mile away, leaning into the road with his right knee bent and the importunate thumb making a rigid point at the far horizon.

He looked so old and ratty I hated to stop. But on the other hand, he looked so old and ratty I hated not to. I eased down a hundred yards past him and looked back to see if he had broken into the hitchhiker's triumphant gallop. He had not. He did not believe that I or anybody else would stop for him, but he had that knee bent and that thumb up again for another car that looked as big as a June bug far back on the highway. I had to back up and persuade him to get in.

He was one of the most weatherbeaten specimens that ever came out of the brush—an old man bundled in decayed sweaters and defended feebly by a pair of overalls that were in truth on their last legs. About three days' growth of white whisker waved gently over his furrowed features; a collection of fine mahogany teeth peeped out from between his cracked lips; rheumy old eyes lurked behind his steel-rimmed spectacles; and a hat that had known no rest or mercy for many years, high of crown, wavy of brim, and stained with unnamable stains, was jammed down on his head.

Charley is a little deaf, and you have to holler at him, as I found when he didn't hear my friendly inquiry after his health. He went right on, aloud, with what he had been thinking.

"I been standing there two hours," he said. "I coulda been in Seymour and half through with the dentist by now. I was just about to turn around and go home and catch the bus to Wichita Falls tomorrow.

"Yes, I live here. My son-in-law sort of takes care of things on the Wilson ranch over there. I worked on ranches all my life myself—between here and Fort Worth."

"What kind of jobs did you have?" I bellowed at him.

"Just an ordinary hand. I never had sense enough to keep anything."

"Who are these Wilsons?"

"Oh, they came here in the early day. Ed Wilson was the old man. He started the ranch. His son Jim Wilson and Jim's son are

115

on the place now, living close together. They ran into oil and have a lot of money. Live in town part of the time.

"No, they ain't changed much—still about the way they always was. Except maybe Jim's son. He drinks from half a gallon to three quarts of whiskey a day. His father drinks, too, but he does his drinking at home and don't bother nobody. Sometimes he plays drunk tricks, but he don't mean no harm and nobody minds—except my son-in-law when he's working them fool cows. It's hard enough to make them do what you want without any interference.

"Yes, my son-in-law is a mighty busy man. He takes care of six pastures; just under four hundred cows. He's got no help except at branding time. He needs some, too, specially in wet weather when you can hardly get in to the cows. Them big oil trucks cut the pastures up so bad you can get stuck anywhere. Ought to have two men all the time.

"The cattle are all Herefords. No, no Brahmas. Now on the Waggoner ranch they got a few Brahmas, and they're so wild you nearly have to tie them to the fence before you can rope them. I don't like them critters. They'll git in yore britches. You go to messin' around with them things and they'll git in yore britches.

"How's that?—How do we live? Oh, all right, I guess. Not as good as we used to. Folks don't get so much meat now. The boss man has it—keeps it in his locker in town. Poor folks have to buy it, and they don't buy much. Mostly they eat beans and bread and milk."

I looked at the rolling sea of mesquite brush and thorny bushes and asked if he remembered when the country was open. "Yes," he said, his old eyes far away. "This brush has grown up in the last thirty years. Everything was open when I was a young man, except for some big mesquite trees. The old longhorns got around all right because there wasn't so much brush to catch on. But you could hear their horns a-knockin'— a-knockin'—against the branches."

Seymour was just over the hill. I asked him one more question: "Do you ever hear the cowboys sing any more, or do they turn on the radio?"

"No," he said. "You don't hear 'em now. But when I was young in the early day you could hear them in all directions— a-ridin' and a-singin'—a-ridin' and a-singin' . . . ."

I let him off on the northwest corner of the square at Seymour, and he took out on a diagonal for a second-floor dentist's office. "Thanks for the lift," he said over his shoulder. "I sure appreciate it."

PART III It Takes All Kinds

I 10. Mexican Style

IN THE FALL of the year 1896 a very important football game was played at Lawrence, Kansas. The invincible Kansas University team was playing the unbeatable eleven from the University of Michigan. Each side was supported by droves of noisy rooters. Insults and money flowed freely. By kickoff time several thousand people had jammed into the bleachers, all in a high state of tension. A visitor from another world would have thought the fate of Christendom hung on the outcome of this game.

On the sidelines in a steaming frenzy of excitement was a short, stocky, intense youngster from the Lower Río Grande country in Texas. He had black hair, a smooth, round brown face, one front tooth out of line, and a Mexican accent as thick as *sopa de arroz*. His temperature was at least three degrees higher than any one else's that sunshiny fall day; and when the invincible Kansans proved to be less than invincible, he took it hard.

The youth was Joe Canales, now a prominent lawyer of Brownsville, Texas, and his story points up some interesting contrasts and developments involving new and old-style Mexican ranchers. Joe belongs to the third generation of a notable ranching family. The fifth generation is just growing up now.

At the time of his great disappointment Joe was about to graduate from high school in Kansas City, where he had been absorbing knowledge at the behest of his grandfather. He had not particularly wanted to come. In his inmost heart he wanted to be an engineer—he loved figures—or even a doctor. The last thing he would have chosen for himself was the fate of a lawyer sweating over hard cases in a musty courtroom. But don't think he ever said as much to Grandfather. Children of Mexican stock

121

in those days were brought up to do as they were told. Even now they don't have as much to say as *Americano* children when an edict comes down from on high.

Grandfather needed a lawyer. He had sent one of his sons to Monterrey in Mexico to prepare himself, but this son had been seduced from his path by the social attractions of that delightful town. So the old man had legally adopted Joe as his own son and sent him off to learn the ways of the gringo and the laws of the gringo in the land of the gringo.

He *did* need a lawyer badly.

Like many of the Mexicans who settled in southern Texas, Señor Canales was a native of the town of Mier in the state of Tamaulipas, and was as much at home on one side of the border as on the other. After the revolution of 1836 he joined the exodus of Mexican citizens of Texas who hoped that life would be more peaceful if they put a river between themselves and the rampaging Texans. Some tried to drive their herds before them, but it was necessary to hurry, and it is impossible to rush a cow—unless you want her to go slow. So the country became a refuge for escaped cattle and their wild, unbranded offspring.

The next few years were a bad time—a very bad time. Mexican bandits raided, stole, and burned. Then the Texans raided, stole, and burned in revenge. The few Mexican ranchers who held out through those terrible years had haciendas built like forts, and bands of retainers who were hired to fight as well as work cattle. Veterans of the Mexican War settled in the region and feuded with their Latin neighbors. Former Confederate soldiers came in after the Civil War hoping to get a start by branding the unclaimed cattle of the trans-Nueces region. If it wasn't one thing, it was another. Nevertheless, there was a tendency for Mexican families with sufficient courage to drift back into the territory after the Mexican-Texan business had cooled off. Señor Canales was one who had the courage.

Like many others, he was a sheepman when he recrossed the Río Grande, but he soon switched to cattle. There was some pressure on him to do so (threats were directed at him and other sheepmen, says Joe Canales), but he was ready and willing to change for the sake of keeping up the pastures.

His major trouble was not sheep, but lawyers. He wanted to add to his holdings and got a good price on the estate of a deceased Mexican named Juan Moreno. The heirs lived in Mexico, and were eager to sell, so Canales bought the land and settled back to enjoy his possessions.

Then he found that he had overlooked something. In Mexico a husband can act as his wife's agent, and if he signs a deed in her name the transaction will stand up. In the United States, with its peculiar habit of treating women as if they had sense and responsibility, the wife can raise a successful protest if she hasn't been consulted. Consequently a number of Moreno heirs hired American lawyers and sued to declare the Canales title void on the ground of not having given consent.

The lawyers could have gone down into Mexico, assembled the clan, paid off everybody in a communal palm-greasing ceremony, and settled the business once and for all. But that would have been too easy and much less profitable. They preferred to deal with the objectors one at a time in trial after trial, and this individual treatment nearly broke Grandfather Canales. To keep from being bankrupted, he decided to make a lawyer out of one of the family. Surely one of his own blood would not rob him. Anyway, he had better not try.

That was why grandson Joe went to Kansas City. He was all ready to begin his legal studies at the University of Virginia when the Michigan University team achieved the impossible. So impressed was Joe with their performance that he bought a ticket to Ann Arbor instead of Charlottesville and the sheepskin that hangs on the wall in his new office in Brownsville bears the seal of Michigan.

It is sad to relate that Joe got himself admitted to the Texas bar only in time to help his grandfather through bankruptcy proceedings. He was able to save the homestead, and that was about all.

Later on, after he had been admitted to practice before the state Supreme Court and had become a member of the legislature, he finished the Moreno business once and for all by asking the state Land Office to institute a test suit to determine title. The Commissioner looked at the papers Joe had accumulated

123

and shook his head. "No, no," he said. "You'd beat us." And he made notations in his records which thereafter kept poachers off the Canales preserves. Thanks to this move, the family stayed in the cattle business.

Joe, however, was never the rancher of the family. His brother, Don Alvino Canales, is the man who stuck to the cows and brought up his sons to do the same. He may well stand as a model for the best type of third-generation Mexican rancher. He is a strong man in the best sense of the term—a political leader, a peace officer, a protector of his people. The patriarchal system which is the basis of life in rural Mexico and in the most Mexican parts of the United States finds its best justification in men like him. Judge Harbert Davenport of Brownsville says that Alvino Canales is "without exception the best man and the best citizen I know."

Don Alvino is a cattleman to the backbone, too, but not in the American tradition. How could he be, having seen what he has seen and remembering what he remembers?

He remembers the open prairie, the tremendous ocean of grass that waved over uncounted miles in the Texas of fifty years ago—"so open you had to dig a hole and tamp the bridle rein in to tie a horse." It is a forest of mesquite now. He remembers the great drift of unclaimed cattle which filtered down ahead of the sweeping northers—the wild cowboys, ignorant, violent, and untamed, who hunted cows as if they were game animals. He remembers the awful drought of 1893 when the seventy-foot well on Grandfather's ranch went dry and the stock died of thirst, panting and moaning—the change-over from sheep to cattle at the end of the eighties—the gradual alteration of centuries-old Mexican ways and even the disappearance of the old-style houses made of mesquite posts set on end, plastered, white-washed, and roofed with bundles of native grass laid like shingles. First people had to have wood floors when cheap lumber became available. Then they began to feel that it was too much trouble to build the traditional houses at all, and "American houses" sprang up, hot in summer and cold in winter. In one of the Mexican houses the temperature could change thirty degrees outside and the inmates never knew the difference.

I first saw Don Alvino in the Nix Hospital in San Antonio, where he was unwillingly spending a month in bed. He is no limber youth any more, but until recently he refused to allow his age to get in his way. Every day, and sometimes two or three times a day, he would drive out to the little ranch he still runs six miles outside of Premont (one of several that he owns) to look at his cattle and see how his *caporal*, Rafael García, was doing. One day he started to climb down off a corral fence, slipped, caught his boot between the rails, and nearly jerked his foot off. Infection set in and they told him he would have to lose the foot.

"No," he said. "I will not have it cut off. I would rather not go on than to live a cripple."

Under protest, the doctors tried rebuilding the tissue with dried red blood corpuscles, and succeeded.

So there he lay in room 1918 in a pair of blue-and-white-striped pajamas, a noble-looking Mexican with the face of a kindly old lion, a shock of white hair, and a pair of steel-rimmed bifocals, full of memories of days gone by.

He came to the portion of the family property called La Cabra Ranch, west of Corpus Christi and north of Falfurrias, in 1879. At that time La Cabra included only 160 acres and was a long way from anywhere. Towns like Alice were far in the future, but the road from the valley settlements to Corpus Christi ran past the door. For those times the family was well off. Alvino's mother came from rich people and brought to her marriage eight cows and fifty horses. His father had a number of sheep. Before long they began buying land, and this, too, at a time when a man was regarded as mentally unstable if he thought it necessary to own land when practically the whole western part of the continent could be used free. Nevertheless, when the old man died, he had 40,000 acres. Ranch land brought one to two dollars an acre when he started buying.

It was a struggle, of course, to stay alive and still pay for all that property, but the family had one stroke of luck. The drought of 1893 wiped practically everybody out. It left Alvino's father with only 193 head of cattle. But fortunately he had sold his fat steers just before the calamity struck, and consequently he had the money to rebuild when the rains came again.

Times have never been that bad since, but as Alvino grew up and assumed more responsibility, it seemed as if he were always fighting for something or against something. Fortunately he was a good fighter.

For two years during the first world war he was in the Ranger service chasing thieves and slackers. Already he had begun to take a rather discouraged view of human nature. "There are more bad men than good men," he says. "You know that yourself."

Human nature seemed to get worse rather than better after the war. Thieves ran off his stock almost as fast as he could raise it. Finally he told his men not to dillydally with anybody they caught red handed. As a result a couple of rustlers got killed, and Alvino was in difficulties. One of his cowboys on the Hebronville ranch was arrested and charged with murder. Alvino was prepared to go the limit in defending his man. "I would rather spend a thousand dollars than have the stealing go on," he said.

The case was moved to Jim Hogg County, famous in more recent years for political machinations. That was all right with Alvino. The District Attorney challenged every cattleman on the jury panel. Pressure was brought on Alvino to get him to take out stock in a certain company in return for some back-door influence. He didn't want any of that stock.

"We'll send your man to the penitentiary," he was told.

"All right," he replied—"if you can."

The man was acquitted. The dead rustler had cut him with a knife all the way down one side. It was a clear case of self-defense.

It was now time for Alvino to see what he could do in public office. He would not have been a Canales if he hadn't got his name in the ballot box sooner or later. He was county commissioner in 1924 when the first county highway went through—from Alice to Falfurrias. With a few such achievements behind him, he began to swing a good deal of weight in the county. Then the propositions began to come in. One man wanted to be county judge and asked for his support. "You elect me," he said, "and you'll be the judge."

"I didn't like that," says Alvino. "I defeated him, too."

CATTLEMAN'S MORNING
A cowboy "busts" one on the S.M.S. in Texas

As a sort of climax to his modest political career, Alvino is, at this writing, mayor of his home town of Premont.

Bad days came for the last time in 1933. Captain Jones asked him to come to Alice. He went—and was told that the bank in which he was interested would have to be closed. "I want you to tell everybody they will get their money," Jones told him.

Alvino worked hard at it, and his friends and neighbors trusted him. The bank stayed closed seventeen days; then reopened and paid off 25 per cent. Ninety days later it paid off another 25 per cent, and eventually paid in full.

Soon afterward men with derricks and pipes brought the black magic of oil out of the ground, and there were no more worries about money in the Canales family. On the three thousand acres of La Cabra, managed now by Alvino's brother Praxedes, there are ninety-four oil wells. The rest of the property has almost as many.

Don Alvino is not so fond of the wells. He thinks they spoil a ranch. They have certainly added to his problems and taken the freedom out of ranching. A dozen different kinds of people have business inside his fence. Two families of oil-company employees live on the place he looks after personally, and he has a chain on his front gate with eight separate padlocks on it, each padlock serving as a link in the chain. A man who has a key to one of the locks can enter the place. Everybody else had better stay out.

The fourth generation is represented by Alvino's four children: Miss Tomasita Canales; Mrs. Chris Hinojosa; Gus, who mixes ranching with oil-field construction; and Mrs. Charles Hornsby, whose husband is in business with Gus.

Gus has a story to tell, too, and is a living example of how the old Mexican families can forge ahead in the land of the gringo. He is a short, boyish, quiet fellow with a great bush of curly black hair just beginning to show a little gray. He is so nearsighted he has to have a chauffeur to drive him around, but there is nothing wrong with his mind, or with his courage.

When he got married, to an Oklahoma girl of German ancestry, Alvino told him he would have to get busy and make a living—as if Gus needed to be told. He started in with the Ford

127

agency in Falfurrias and soon became sales manager. But that wasn't good enough. The real money in those parts was in oil, and he wanted to get closer to the oil business. Refusing to borrow a cent from his father, he put out $500 as down payment on a bulldozer and went into construction.

Don Alvino thought he would never make it. The payments on the bulldozer and other equipment totaled $800 a month, and where was it going to come from?

It came without too much trouble, thanks to the contacts Gus had made in the automobile business, and at this moment Gus Canales, Inc., owns a hundred thousand dollars' worth of machinery. The firm keeps three hundred people on the pay roll, operates a private plane (partner Charley Hornsby was formerly in the Air Force), and does a tremendous volume of business. Gus works in the office and lines up the jobs. Hornsby, a slender, relaxed South Texan who could pass for a cowboy if he had a little more sand in his ears, handles the crews in the field.

But this would not be the story of a great-grandson of old Señor Canales, who came out from Mier one hundred years ago, without some more politics. Gus has been right in the thick of a political fight, and against the most powerful foe he could find —a notorious political machine with tentacles which reach out all over South Texas. For years the leaders of this ring have done as they pleased in their own territory, employing whatever tools and methods came to hand. In one respect they are no different from other politicians: they want to control everything they can get their hands on and then reach a little farther to control everything else. Jim Wells County is not part of their home territory, but they keep trying to take it over.

For eight years, from 1940 to 1948, Gus was county commissioner from his precinct. Every two years, at election time, there was a fight to the finish. The Canales clan is large, and members always took part on both sides. Consequently much heat was generated inside the family at each election, and each time it took a few months for it to cool down.

In 1946, Gus almost lost out, but his shrewd foresight and precautionary measures, together with his own personal brand of courage, had good effect, and he won. In 1948, however, the story was different. Although there were no fireworks, the plans of the enemy were evidently deeply laid. And although it appeared that he had been elected and the Corpus Christi *Caller* came out with that announcement next morning, two hundred unexpected votes were found in one of the ballot boxes, and Gus was out of politics.

When the smoke cleared away, Gus turned up what seemed an amazing mess of skulduggery. F.B.I. men and postal inspectors investigated alleged irregularities in connection with absentee ballots, but for some unexplainable reason did nothing at all. Gus took one case to the Supreme Court of Texas and won it, but with the authorities seemingly completely lacking in interest, nothing further has been done.

Gus says now that he is almost glad he was beaten, though he hates to see those fellows get away with it. A county commissioner in South Texas has just too many people and responsibilities to take care of, not to mention the peril to life and limb.

It will take one thing more to make Gus perfectly contented. He wants to get back on a ranch. With four generations of ranchers behind him, he can't get the business out of his blood. He and his partner Charley have bought nine thousand acres on the Nueces River and plan eventually to turn the construction company over to somebody else so they can give their whole attention to ranching. They both feel that their children would be better off in the country, and would like to see the fifth generation of the family learning to be at home in the saddle.

Meanwhile Gus works night and day, eats flour *tortillas* and beans, consumes fourteen to twenty cups of coffee a day with a quart of milk for contrast, and wonders what will happen in the next election.

11. Foreigners

SOME DAY I hope to see a Western movie with a hero named Hasenpfeffer, but I expect to have to wait a long time. In books and moving pictures we have built up a myth which keeps everybody but "old American" families with British names out of the cattle business. A cowboy named Le Beau Soleil or Kothmann or Dvernik would be unthinkable.

In life, fortunately, no such taboo exists. It is surprising how many people from the far-away places have brought their strange-sounding names to the cattle country. Take the first one we mentioned. Le Beau Soleil was the right name of a celebrated Wyoming character who came out in translation as plain Tom Sun. He was a French-Canadian orphan who made his way to Vermont, to St. Louis, and finally to Wyoming long before the Civil War, and became a government scout and Indian fighter— a mighty man with a rifle. When he settled down, he picked a place near Inscription Rock on the Oregon Trail. The Sweetwater races along past the ranch house he built in 1872 at the foot of a massive upthrust of volcanic rock, stained with water, seamed with lichen, and freckled with little pines and cedars.

Tom's son, also Tom Sun, runs the ranch now, and likes to show visitors the .44-caliber Evans rifle, dated 1868, which Buffalo Bill presented to his father. He has a museum full of relics of the earlier days, including a photograph of the elder Tom in his best clothes—a tall, heavy-haired, ramrod-straight figure with a pair of intense, almost hypnotic, eyes. The neighbors say Tom wasn't much on reading and writing, but he knew Indians and cattle and was very much a man.

Tom Sun's son Tom is pure Wyoming, with no trace of his French ancestry left. Everything about him is relaxed. He sits as if he were lying down and walks as if he had just oiled his knees. Even his shirttail pauses halfway out of his pants as if it were too much trouble to climb any farther. He wears a great big hat which he sometimes lifts a little to push back his thick gray hair. Otherwise it stays on his head, and he probably sleeps in it. The top half of him is covered with a khaki shirt, and the bottom with yellow corduroys half in and half out of his boots. He talks with a minimum of mouth movement, but a small smile hovers all the time about his short upper lip. He hardly needs to mention that his mother was an Irish girl from the Old Country, bright as a dollar and peppery as a wild onion.

All the north country is full of French memories. There was

Pierre Wibaux in Montana and the Dakotas who left his name on a cow town; there was the Marquis de Mores whose dreams of a ranch empire, complete with packing plant, private stage line, and personal palace, make good story telling at Medora; there was Peter Duhamel who became a cattle king in the Black Hills Country.

Duhamel wrote down his story in his old age for the sake of his grandchildren. It is a hair-raising tale about another French-Canadian boy whose experiences were probably not much worse than those of a hundred foreigners, now forgotten.

Peter left home at the age of eighteen when his brothers allowed him to take his small share of the family inheritance and strike out for the American West. In the spring of 1858 he trekked out of St. Louis behind a pair of mules belonging to a squaw man named Louis Regard. The expedition was on its way to the Black Hills, then the wildest sort of Indian country, to trade the Sioux Indians out of what furs they had. It was a crazy thing to attempt, and nobody was very hopeful about getting back alive.

Regard was a shaggy brute who made his men live on hardtack and coffee and infrequent baits of wild game. There was only one gun in the party. Regard carried that himself and allowed no one else to touch it. Once when they were all about to starve, he shot an elk cow, but the men never smelled a steak. Regard's squaw, who did the cooking, knew they were hungry enough to eat anything, so she began her task by popping all the entrails into the pot and inviting those interested to help themselves. Peter reached in, got a leg of the elk's unborn offspring, and wolfed it down.

They found the Indians, and the Indians were hungry, too. As a result, Regard's stinginess bore fruit. All the supplies he had withheld from his men now went for furs, and soon he had ten wagonloads of pelts—more than he could haul. Somebody had to stay with the excess baggage while Regard went back for more wagons. "It was decided by chance," Peter tells us—and Peter got the short stick. As the wagons creaked away, he stood in the middle of the prairie, all alone, with no provisions beyond a little hardtack and coffee, a buffalo-hide tepee, and a few wolf traps, and watched the rest go with feelings that can hardly be imagined.

"For two months and eighteen days I lived alone in this manner," he says. "My hardtack and coffee I reduced to quarter rations from the start and continued in this manner until the last morsel had been absorbed. At the beginning of the fifth week I was subsisting entirely upon the meat from the wolves I could catch in my traps, and continued to subsist in this manner until the train returned. How often I wailed aloud no human soul but mine can tell, and how I managed to cling to life through those terrible lonesome days and weeks I cannot now explain."

But cling to life he did. Eventually Peter Duhamel became one of the richest men in western South Dakota and left his memory and his sturdy fiber to a flock of descendants. One of them, Emmet Horgan of Rapid City, is still carrying on in the cattle business and is now the president of the South Dakota Cattlegrowers Association.[1]

Yes, those dogged youngsters, sometimes without a word of English at the time of their arrival, did their part in building up the cattle country—or tearing it down, as it seems to be the fashion now to say. The annals of the range are full of accounts of foreigners like Conrad Kohrs and Johnny Bielenberg, Germans, who became cattle kings in Montana; like the Kothmanns, Geistweidts, Brandenburgs, and Lehmbergs, who still do business in Texas; like the Broussards, Landrys, and Arceneaux who came out of Louisiana long ago and raised cattle along the Gulf of Mexico.

Their descendants often keep some of the old flavor. A prominent rancher in western North Dakota is Anders Madsen, whose father left Denmark at the age of nineteen and headed for the wilds. He worked with a gravel train near Medora until he was twenty-one, then filed on a homestead and set up for himself. He became a cattle raiser, and made cattle raisers out of his sons.

Anders is a short-legged, barrel-bodied fellow with a face like that of a friendly Brownie. He looks like the son of a Dane,

[1] W. J. Todd, "Peter Duhamel the Pioneer," MS in the possession of J. E. Horgan.

and still speaks with a bit of an accent, but he talks like a cow-boy. "Hell is full of cowhands that tried to brand with a cold iron," he remarks as he gets ready to work on some yearling bulls that Ag Kennedy has just come for. He thinks like a cowboy, too. His son-in-law takes care of the feeding—Anders says he can't do it himself. "I used to sweep the granary floor for horse feed, and I just can't throw the stuff at them like you have to do to put on the gain."

Another Scandinavian rancher lives just outside of Walden, Colorado, capital of the lovely, mountain-circled meadows of the North Park. Fifty years ago he was a tow-headed Swedish boy hanging over the rail of a transatlantic liner watching the stern heave and dip as the mid-ocean rollers raced forward along the wet black sides. Sixteen years old he was, and most of the time during those sixteen years he had dreamed about being a cowboy. His friends said that the middle initial in his name— Carl D. Johanson—stood for Diamond Dick. He had given his parents such a hard time over his ambitions that they finally weakened and sent him to an uncle who lived in Colorado.

The next move (in case another ambitious Swedish boy wants to know how it is done) was to change his name to Carl Johnson and take a course in a Denver business college. That kept him busy and improved his English, but he never had any intention of being a city boy. As soon as he was old enough, he filed on a homestead. He worked hard at improving his property, but there came a time when he had to register his holdings in order to stay in possession. Somebody else had his eye on the claim, too, and almost beat Carl to town. But this snake in the grass had not reckoned with Old Jasper, Carl's white horse, the fastest thing on four legs in North Park. Jasper got his master there first, and that is one reason he is running a 125,000-acre ranch now.

You should see that old Scandinavian in boots and big hat loafing along on horseback. If you don't think a Swede can make a cattleman, just look at him.

"Well, maybe a Swede," you may say, "but I'll bet you never heard of a Czech in the cattle business."

Adolf Adamcik of Smithville, Texas, can take care of that

133

statement. Adolf is a Czech, a good citizen, and a cattleman through and through. "I'll be a cattleman till I die," he says. "If I had to sell everything, I'd be the most miserable man you ever saw. I'd as soon go to jail as sell out. I want some cattle around just to look at."

Grandson of a Bohemian immigrant farmer, Adolf was brought up behind a plow, but when he was fourteen, he picked cotton for a neighbor at thirty cents a hundred and bought a cow with his earnings. "I've never been without a cow since," he says. But one cow does not make a cattleman, and he still had some plowing to do.

"After a few years my father bought 500 acres just south of Smithville. We were the first Czechs to move in here. I wanted cattle, but we had 125 acres of farmland and that had to be worked, too. My brother and I used to plow by moonlight till midnight so we could work cattle the next day."

In 1910 he had his chance. A Mr. Marburger in the neighborhood wanted to move to San Antonio and had to get rid of his stock. He offered thirty-five steers to the boys for twenty-seven dollars each. Their father was dubious, but their mother was all for it. "They'll never learn if they don't take a chance," was her opinion.

The boys bought the stock without asking their father to go on their note. In the spring they sold out for sixty dollars a head.

It was up and down after that. In 1917, Adolf opened a meat market in Smithville to sell meat which was not moving on the hoof. One of his customers was a Scotch-Irish schoolteacher who liked the meat well enough to marry the butcher. She is a hard-driving field worker for the Baptists, and Adolf is a Roman Catholic, but she respects his rugged virtues, and he values her enthusiasm and high-mindedness.

Nineteen twenty-one and 1934 were the years when he hit bottom. He still shudders when he remembers the bags of bones that used to be fat cattle stumbling from the brush to the water tanks in thirty-four. Some of them were so weak that when they lay down after drinking, the extra weight of the water made it impossible for them to get up. "If it hadn't been for government help, I never would have come through," he declares solemnly.

Now he runs between six and seven hundred head of cows in ten big pastures and some smaller ones. He makes the rounds in a Model A Ford, charging furiously through the brush, making entries in the paper-covered notebook which is his entire bookkeeping system. He runs three brands and knows every animal he owns, as a good cattleman has to do.

With his background Adolf is naturally no cavalier of the plains to look at. He is a sturdy, gray-haired citizen with a weather-bitten face punctuated by a couple of missing teeth. One eye is out—the result of a raid on his father's trunk when he was six years old. He found some percussion caps, exploded one against a rock with a hammer, and paid the penalty. Now he wears a pair of old-fashioned gold-rimmed spectacles with a dark lens over the missing eye and no lens at all on the other side. When he picks up a magazine, he puts on a pair of reading glasses over the original pair. His ordinary costume is khaki shirt and pants over a pair of low-heeled boots.

Beneath this unpretentious exterior are many of the old-country virtues so often lost in the process of Americanization. Adolf is a grim worker, keeps everything he has in top condition, and is utterly and completely honest. He does not try to find exemptions and little dodges to cut down his income tax, for instance. If he owes it, he says, he wants to pay it.

A similar tradition produced Jack Dvernik, a Russian immigrant who owns most of the land around the Killdeer Mountains on the wide grasslands of western North Dakota. Forty years ago he was running his legs off as a waiter at the St. Charles Hotel in Dickinson. He got a little ahead and bought some land which made money when he planted it to flax during World War I, raised a bumper crop, and collected five dollars a bushel. The next year he traded for more productive land and did even better raising wheat. He bought more property, made more money, and proved to be the only man in those parts who had the $22,000 necessary to buy the old Diamond C when it came on the market. He moved out to the headquarters twelve miles from the little cattle town of Killdeer and began running his legs off after cows instead of customers.

135

The ten bad years which followed were a blessing to Dvernik. Others went broke—and he bought up their land. But he never lost his balance. Even now he admits that everything he knows about cows he learned from his neighbors.

His Russian heritage is still strong in him. A solid man with a close-cropped mustache, he walks with an erect carriage which goes back to his days in the Czar's army. He lives very simply in a plain house, but has a wonderful orchard and keeps his ranch up as a racing driver keeps up his car. Two broods of grandchildren, his own and his second wife's, he cherishes with patriarchal enjoyment. He is deeply religious. Old-World courtesy and Western hospitality are joined together in him, but when he invites a traveler in to dinner, he asks him gently not to smoke in his house. He does not believe in smoking.

Almost any country one could name, with the possible exception of Lower Slobbovia, has contributed a son or two to the American cattle business. Scotsmen like Henry Grierson of Hysham, Montana, are not at all uncommon. Henry came over with his family in 1881—long enough ago for him to lose his accent but not long enough to take away his craggy Scottish personality.

Noble Englishmen, like the Frewens and the Aylesfords of former times, are all gone now, but M. H. W. (Monty) Ritchie came over from Ireland to take charge of his inheritance, the J. A. Ranch in the Texas Panhandle, in 1932. Well educated and used to a life of aristocratic luxury, he faced the facts, put on some work clothes, and buckled in. With over 400,000 acres on his hands he knew the armchair was not for him. He assembled a new staff, hired Bill Word as manager, reorganized the work by putting special crews at special tasks, and tightened up wherever he could. To make sure the job was being done, he went out with the wagon and did his share of the work. The men like him in spite of his English accent and his meticulous management— no small tribute in a country where they still tell stories illustrating the gullibility and inefficiency of Right Honorable cattlemen of sixty years ago.

At least one Austrian is in the business, and it would pay any student of American culture to drive fourteen miles east of Castle Rock, Colorado, to visit Josef Winkler and see what he has done.

Don't go in bad weather, however. The foothills in which Josef lives are up close to seven thousand feet—two thousand feet higher than Denver, which is only forty miles away, and the roads can be as slippery as a Chicago politician. The hills are high and steep, clothed with brush and scattered trees; the valleys are deep and sinuous with occasional broad bottom lands where the hay grows tall and the small grain is thick.

The Winkler place is down in the valley of Cherry Creek, its center the big square house of concrete blocks with a two-story frame addition at the back. Barns, corrals, and shops stand aside at a comfortable distance. The tree-studded creek bottom lies beyond, rolling fields rise on the other side, and steeply sloping pastures climb up to shut off the horizon still farther away.

Inside is solid simplicity with no frills. The kitchen at the back of the house is a huge place as full of machinery and gadgets as the nerve center of a beef factory ought to be. The dining room is battered by decades of hearty feeding. The living room is small, furnished with sofa and chairs in the best Sears Roebuck taste, and obviously not much used except for company. The only decorations are the glittering silver cups, trays, and services which have been won by Winkler's prize Shorthorns.

Josef will sit you down on the scarlet plush of that sofa and give you as much time as you need. As he comes in, you feel the European in him. There is a suggestion of refinement in the way he carries his big body, walking slightly stooped with his knees snapping out; in the sweeping gestures of his hands when he warms up in conversation; in the little extra grain of courtesy.

He does not try to look like a cowboy. In fact, he says he doesn't like this "cowboy stuff" and tries to keep it out of the pattern in his family. His face is as wind-reddened under his light blond hair as if he lived in the saddle, but he wears cowboy boots only when he has to ride a horse, which is not often. Bib overalls and rubber boots feel better to him.

A rich Central-European accent would give him away, if nothing else did. A storm is still a "sturm" to him. When he remembers his early days in this country, he says, "I never had no fun. When the others went to dances, I stood at home and took care of the cows."

137

It is strange to see this European in the midst of his thoroughly American sons and daughters. Can he be the father of this brood, you wonder. When you know him better, you wonder if the country wouldn't be far ahead with a few thousand more iron-willed, inexhaustible Josef Winklers.

It was a fight for Josef from the beginning. Everything he has achieved has come the hard way. His father was a cattleman and farmer who raised a special breed of red cow in the mountain pastures near Linz and the Brenner Pass. Josef's cow education began early as a result of the local custom which sent all the children up into the high country with the cattle every summer. They stayed alone for months at a time, doing their own cooking, enduring week-long storms, and staying with the job as a matter of course.

There were gay parties in the winter (Josef played the harmonica), but life was full of hard work, education was hard to come by, and the future was not promising in Austria. The most illuminating thing in Josef's life was the fact that he had an aunt in Colorado who was married to a cattle grower named George Engel. There were letters telling how Engel, a Denver carpenter, homesteaded east of Castle Rock, fought the Indians, went to the legislature, and became a prominent man; how his first wife became ill and Josef's aunt left her job as a domestic servant to nurse the sick woman; and how she married Engel after he became a widower. Josef ate those letters up.

Finally it seemed to him that he had to go to America. Why should he, a younger son, stay on in the Old Country with no prospects except unending service in the Austrian army—three years' enlistment, three years' maneuvers, and the rest of his life on call? But his father would not let him go. It took much stealthy work on the part of a brother plus some encouragement from the mayor of his village before he made it across the border into Switzerland. After that it was easy.

His aunt, now married to Charles Breuss, was not expecting a visitor when he appeared at her door in May, 1907, but she was glad to see him. They needed help on the ranch.

Breuss was a hard worker and was building up a fine stock farm, but the stock was getting the short end of his interest.

Winkler, on the other hand, was more interested in cattle than anything else in the world. About 1910, Breuss let him take charge of the herd.

If you ask him whether or not he has any special gift with animals, he will modestly admit that he has. "It's something like a natural gift for music," he theorizes. "Some people have it and some don't. I guess you could say that I was a natural judge of livestock—I was sensitive about them. Right now I forget people, but I remember cows. No, nobody showed me how. I always had to work alone. But I went to the stock shows and got a lot of ideas."

One of his ideas nearly shocked Mr. Breuss into temporary paralysis. Josef wanted to improve the Shorthorn herd by getting hold of some better bulls than they had been using. He found one that he wanted in Denver. It cost $600. That sounds like a bargain now, but in those times a man who paid more than $100 for anything on four legs, including a piano, was looked at with sympathetic anxiety. What made it worse was the fact that just about the time Mr. Breuss was becoming reconciled to the hole in his bank account, the bull died of blackleg. Mr. Breuss had not been willing to fool around with new-fangled ideas about vaccination.

It took some time for Josef to live that down, but he held on. His stock improved, and he built up friendly relations with the Weiss brothers at Elizabeth, who taught him a good deal. In return he helped them fit the calves they bought from him, and carried off the Grand Championship in the Denver show. From that time on Joe Winkler was in the show business—the trickiest, most unpredictable route a cattleman can take. He bought the G. E. ranch—3,000 acres and 400 Shorthorns—in 1919, and won his own first Grand Championship at Denver in 1925.

Still he was not satisfied. The Herefords were forging ahead and the Shorthorns were receding. Everybody seemed to want low-slung blocks of beef with white faces. "Why," Winkler wondered, "couldn't a blockier type of Shorthorn be developed to match those all-conquering Herefords?" In 1929 he decided to try to change his stock to meet the new situation. When he talked about it, however, he ran into a hard wall of incredulous laugh-

139

ter. For once he didn't go ahead, and it is the one thing in his life that he looks back on now with shame and regret.

"I didn't have the courage," he told Willard Simms of the *Record-Stockman* twenty years later.[2]

But he got his courage back in a few years and began the long search for the ideal bull—the cattleman's pot of gold. The result was apparent in 1944 when he had the Reserve Champion at Denver—in 1945 when he did it again—in 1946 when he reached the mountain top with the Grand Champion lot of feeder cattle.

Albert Fritzler of Sterling bought that lot of winning Shorthorns and took them to the Chicago show. Winkler went along to help him and was heartbroken when they failed to place. But show business is like that, and win or lose, Josef Winkler has done as much as anybody in America to keep those red and white cattle in the pastures. He still thinks Shorthorns can't be beat for all-round purposes.

It was a shock to him, however, when Willard Simms came from Denver in 1947 to announce that he was to be Man of the Year in Livestock. "Why should I be chosen?" he asked. "Dan Thornton and Bob Lazear are more important and a lot bigger than I am."

"That doesn't matter," Simms told him. "Those fellows are tops, all right, but they had something to start with. You did it all yourself."

For a look at a whole clan of non–Anglo Saxons who have made their mark in the cow business, go down to Texas and pay a visit to the Swensons. Those Swedes have been in the saddle for close to eighty years, and show no signs of dismounting.

They operate out of Stamford, a sort of private town which they have laid out in the middle of one of their big pastures. The old office building of the Swenson Land and Cattle Company is just around the corner from Strauss' store a few steps off the Square. Inside, at the top of a short flight of cluttered steps, there is a right-angled counter with a gate in it. Nobody pays any attention to this barrier—just walks in and sorts his way among the tables and chairs to the man or the office he wants.

[2] 1947 *Annual Edition*, 33 ff.

In the corner office sits A. M. G. Swenson (let's call him Swede—everybody else does). The place is crammed with Swede and furniture. There are two desks on opposite sides of the room and a loaded table in between. Everything is loaded. It would take a month to straighten it all out.

Swede is an unassuming, hearty person, American born but built to Swedish specifications with a big-boned, ungraceful body which once served him very well as a top-ranking football player. He leans forward as he walks, with a considerable quantity of hips following him closely. He has an outdoor complexion but few other marks of the cattlemen. He wears brown oxfords, cord trousers, a tan army shirt, and a brown tie. A little fuzzy hat sits on the back of his head. Shell-rim glasses ornament his big nose. He is forty-nine years old and his hair is nothing to get lost in any more, but he comes of tough stock and may last another hundred years.

The family history begins with another immigrant boy who was willing to spend twenty-five hours a day to get what he wanted. S. M. Swenson landed at Baltimore in the middle eighteen thirties with no money and no experience except a few months spent in a store. This, however, was enough to get him a job with a retail firm. He did well. And before long he was sent on a trading expedition to Texas.

In 1838, when S. M. made his first trip, that was no week-end excursion. Railroads were far in the future. The boat sailed up the Brazos as far as Richmond, and there the real business started. S. M. bought mules and a hack and sold the goods he had brought, traveling all over the parts of Texas that were then settled. Before long he was ready to set up in business for himself, first at Richmond and later, when the capital was moved, in Austin. In time he became the biggest merchant in town, a close friend of Sam Houston's, and a power in the land. Swante Palm, the greatest Swede in Texas, was his uncle, and that undoubtedly helped.

All went well until the Civil War broke out. Like Sam Houston, S. M. was not in favor of secession. Sam was big enough to stand the storm, but Swenson "took to the hills." To be specific, he went to Mexico, and one story says that he managed to get a great quantity of contraband cotton across the border and reaped a small fortune.

Another legend, still told in the family, says that just before he left he converted all his assets into gold, got his family out of the house by sending them on a picnic, and called in a stone mason. Well, his chimney *had* been giving him trouble. But when the fireplace was ready to be put back together again, a metal box of gold coins was cemented in beneath the floor.

S. M. did not come back to dig it up. When the war ended, he was established in New York. Swante Palm officiated at the exhumation, however, and S. M. got his gold. He invested some of it in a sugar plantation and spent the rest of his life at or between points in Sweden, New York, and Louisiana.

In New York he became an investment banker, associated with such men as Frank Vanderlip and Mortimer Schiff. In Louisiana, where he passed his winters, he kept up one palatial home, and in Sweden he had another for summer use.

It was amazing how these Swedes stuck together. Almost every year S. M. would bring back a couple of ambitious Swedish youths who wanted to get a start in the New World. There was no charity involved, for they always paid back what he loaned them. At one time the foreman of every one of the four Swenson ranches was a boy S. M. had brought over from the Old Country. In 1888 he brought back his nephew A. J. Swenson, father of the boys who now operate the property.

A. J. started in Louisiana but came to Texas after a bout with yellow fever. For five or six years he helped with the ranch work, a lonesome young Swede in a strange land, and finally thought he had had enough. Back he went to Sweden, figuring on spending the rest of his life among his old friends and relatives. But two things changed his mind. One was a Swedish girl. The other was the problem of making a living. "There were two or three men for every job," he says. The trail led back to Texas.

In time A. J. Swenson became superintendent of the S. M. S., working with Frank S. Hastings, the general manager, who is still well remembered in Texas as a writer, rancher, and philosopher. Hastings was a colorful figure, as much at home in the Saddle and Sirloin Club in Chicago as on a horse directing a

A. J. Swenson, E. P. Swenson, and A. C. Swenson
talking ranch business

roundup. He was the one who publicized the quality of S. M. S. cattle and built up the famous "mail order" system of marketing. To this day the Swensons cut their calves to a rigid classification and simply load and ship when an order comes in. Buyers have learned that they fare as well as if the Swensons let them come down and make a selection (which they won't).

Hastings died in 1922, leaving A. J. as manager—a position he held with the good will of everybody he knew until failing health compelled him to turn the job over to his sons Bill (now manager) and Swede (assistant manager). A. J. is a feeble old man now, a fine spirit in an almost-worn-out body. He has had a stroke and spends most of his time in bed, but gets up sometimes and sits in an easy chair in his bedroom while his nurse, Miss Carlson, hovers about seeing that his knees are covered. The long beard he used to wear is a gray Van Dyke now. His eyes are tired and heavy. But a warm little smile plays around his mouth as he hunts for the right English word and tries to tell how things were in the early days.

Outside the window a four-year-old boy is whooping it up on a tricycle. A. J. watches him with great interest and concern, finally interrupting the conversation to remark that Smoky is the son of Bruce Swenson, a scion of the New York branch of the family who has come down to Texas to learn the ranch business. Smoky would be A. J.'s first cousin three times removed. Smoky has been in to visit him once, and the old man would obviously like to see more of him but does not expect too much consideration from a four-year-old.

The ranch empire which A. J. Swenson and his sons have managed for almost half a century does not belong to them. It is the property of a corporation started by S. M. Swenson before he left Texas. He bought scrip covering railroad lands, purchased additional land, and eventually had three enormous blocks of real estate which he called the Ellerslie, Flat Top, and Throckmorton ranches.

The years went by. S. M. grew old, and his sons reached middle age. The days of open-range ranching receded into the past, and the country went under fence. E. P. and S. A. Swenson, the sons, wanted to put some wire around their properties, too, but the old man did not like the idea. They kept after him, however, until finally he told them they could go ahead if they wanted to and would pay for the fence. As for him, he was going to retire. In 1880 the new regime came in.

E. P. Swenson was the powerhouse of the family from then on until his death in 1942 when he was well over ninety. He was a careful, canny, hard-driving Swede with a terrific will to dominate and, paradoxically, a wonderfully kind heart. The story is told of him that on one of his inspection trips he got ready to go to bed one night in a cow camp and found himself without a toothbrush. A cowboy heard his disgusted exclamation and offered to lend him one. E. P. Swenson took the toothbrush and a cup of water, went out of camp a little way, wet the brush, and brought it back to the cowboy with thanks. He would not risk hurting the man's feelings though he had no use for his property.

Under E. P.'s control the Swensons acquired a fourth ranch at Paducah and finally, in 1906, organized the corporation which bought the huge and historic Spur Ranch on the high plains at the base of the Texas Panhandle. The stockholders were Vanderlip, Schiff, Emery, and Swenson. Emery owned three-fifths of the stock.

The story of the Spur Ranch is one of the great tales of the range—how this empire of grass was taken over by a bunch of New York bankers—how they developed a town in the middle of their holdings and sold off thousands of acres of farmlands between 1910 and 1915—how they combined the sturdy traditions of the Old West with the grim business methods of Wall Street.

About the time World War II began, the corporation was dissolved. The owners divided the property and the Swenson Land and Cattle Company bought all but Emery's share. The Emery heirs wanted five dollars an acre, and old E. P. wouldn't pay it. He said he would not give money like that for land he was used to getting for two dollars an acre. Now that same land is going for as high as twenty dollars. But nobody foresaw the land boom that followed the war, and the Emery holdings were allowed to remain in possession of the heirs. A. J.'s son Eric Swenson is now in charge of the Emery acreage.

All told, the Swensons now control almost 300,000 acres on

144

the four ranches. It is ranch business with a capital *R*, and very much in the old tradition. The men still rope and flank calves in the open pasture, live in bunkhouses or line camps, and work like the devil. But these Swedes have always combined with their native conservatism a certain shrewd willingness to experiment. Most of their property is under lease to the big oil companies and all the lease money is going back into improvements. Close to 220,000 acres have been cleared of mesquite, for one thing. "It's all got to go," says Swede, looking lovingly at those miles and miles of prostrate trunks where only a couple of years ago there was nothing but impenetrable brush. "We don't burn it or take it off the ground. The grass reseeds underneath where the cattle can't reach it, and pretty soon we'll have the old mesquite grass back again. Look under there. It's coming back already."

The brush may go, but it looks as if the Swedes will go on forever. Back in the days when the country began to go under fence a Swenson relative, a Lutheran pastor, came to E. P. Swenson with an idea. The Swedes near Austin and Round Rock, he said, were getting crowded. The young men were marrying and needing places of their own, but didn't know where to go. How about opening up some of the S. M. S. land for them? E. P. liked the idea. He had already tried to sell some of this land, but it had not moved very fast. Maybe giving a priority to Swedes would help.

It did. Not just the young ones, but many older ones also, took up farms around the community of Ericksdale. They and their descendants are there today. The settlement has always been prosperous, but in recent years oil has made it even more so. An oil derrick stands in the back yard of the gray stone church which is the focal point of Ericksdale. The preacher had a right to half the royalties of this well, but he would not take them. He had a couple of oil-bearing farms of his own and was already as prosperous as any man of God in the country. Undisturbed by wealth, he carries on as usual, helping his flock with their personal problems, encouraging soil conservation, and preaching the same sort of Sunday sermons as formerly.

The Swensons drive out to his church from Stamford every

Sunday, sometimes bringing their lunch baskets for the community picnic which usually follows the morning service. A visitor will not hear much Swedish spoken now, except among the older men and women, but if he stays around long enough he will find that the people still cherish the best traits of their stock and combine them with the finest traditions of the American cattle country.

T 12. Managers

WENTY-THREE YEARS AGO, in 1927, a seventeen-year-old Canadian boy was taking his first long train ride on the New York Central. In his pocket was a passport and what was left of the one hundred dollars he had borrowed to finance the trip. In his head were the odds and ends of a high-school education, a healthy fear of the future, and a consuming ambition. Young Bill Roberts yearned to be an American cattleman and was on his way to the Sni-A-Bar Ranch at Grain Valley, Missouri, to start his career. Nobody on the Sni-A-Bar at that time was even aware of Bill's existence, which was probably just as well. Had his arrival been foreseen, he would undoubtedly have been told to stay away.

Green as he was, he knew enough to get himself a berth in the Pullman. They gave him a lower; and when he undressed for the night, he ran into shoe trouble. There was no place that he could locate for his man-sized oxfords. Finally he stowed them away under the berth.

In the morning he reached for his shoes, and they were gone. A strange pair of shiny oxfords had replaced the worn and dirty objects he had taken off the night before. He put them back quickly for fear the owner might catch him and accuse him of larceny.

But he had to have shoes. He looked again. There was something familiar about them. He tried them on. They were his, all right, and that was his introduction to the custom of shining shoes on trains.

146

Then he had trouble with the conductor. The man told him his ticket was made out to Green Valley, Illinois—not to Grain Valley, Missouri.

"But I don't want to go to Green Valley, Illinois," the boy remonstrated. "I want to go to Missouri."

"All I know is what I see on your ticket," the conductor replied gruffly. "You get off at Green Valley."

Too scared and shy to argue, the most discouraged young cattleman in the world was left on the platform at Green Valley watching the rear end of a departing Pullman.

At the cost of more time, trouble, and money he finally found a train and a conductor willing to take him where he wanted to go, and then came the most ticklish part of the business. He had to confront the manager of the Sni-A-Bar.

"I came here to work for you," he told Jim Napier. "I'll take whatever you can pay me."

Jim may have been unnerved by the audacity of it. He gave the boy a job at fifty dollars a month—twice what he could have earned in Canada. Full of joy and thanksgiving, he picked up a pitchfork and went to spreading straw. He was starting at the bottom, but he knew there was room at the top if a man could learn enough.

The pursuit of more and more knowledge has been the key to a successful career for Bill Roberts. Even when he was a boy in grade school at Moffatt, Ontario, he knew he wanted to learn about livestock. On holidays and week ends he hung around the barns and pastures, looked at the cattle, and tried to plan for the future. On other days he studied the stockmen's journals and planned some more.

In the *Shorthorn World* he read about stock farms and cattle ranches—big spreads with fine herds and wonderful breeding programs. Finally he could stand it no longer. He went to his friend T. C. Amos, then and now a prominent Shorthorn breeder, and left with a loan of one hundred dollars to pay his expenses to the promised land.

He probably had little trouble in persuading Mr. Amos to trust him. Bill inspires confidence with his big body, earnest eyes, and slow smile. He is modest without being shy—confident

147

without being brash. Besides he has always known what he wanted to do, and that makes a difference.

On the Sni-A-Bar he absorbed information as relentlessly as a mortgage eating into a ranchman's bank account. The man who taught him the most was Everett Jones, an expert from the University of Missouri in charge of experimental work on the ranch. When Jones had to give up his place after two years, he recommended Roberts as his successor, feeling that Bill had a real flair for the work and a fierce desire to learn.

They turned him down because he had no college degree.

It was a bitter blow, but Bill was not unprepared for it. He had known all along that since his parents couldn't afford to send him to college, he would have to get what he wanted the hard way. He realized more clearly than ever now that he would have to dig harder than other people and stuff his mind till he stretched the seams. He began to look around for a big place with a diversified program which would give him the unaccredited course work he needed. He found his substitute for an agricultural college at W. H. Minor's Heart's Delight Farm near Chazy, New York.

Minor was an amazing character who impressed young Bill Roberts deeply. He had made a fortune in railroading, and another from inventions for improving railroad service. In 1929 his income was nine million dollars. A lot of it went into his silver-plated, gilt-edged, diamond-studded farm.

He had two breeds of beef cattle and two of dairy cattle. He had light horses and draft horses. He had two breeds of sheep and two of hogs. Buffalo and elk grazed his pastures, along with other kinds of game. There was a private fish hatchery. A huge flock of turkeys provided Thanksgiving remembrances for his friends (no turkey was ever sold), and the crates for shipping them were made right on the farm. Once a year some of the buffalo were butchered to provide a princely entertainment for Mr. Minor's guests—whoever they happened to be at the moment.

It took a thousand men to run the place, and there were almost that many kinds of jobs. Opportunities to learn were unlimited, and Bill went through the blacksmith shop, the car-

148

penter shop, the packing house, the electrical department, and a dozen other nerve centers before he had to move on again.

This time it was not his fault. Mr. Minor died on the operating table during a tonsillectomy; the farm was turned over to a board of directors who were mostly bankers; and Bill thought he had better make a change.

The new job was with Doc and Les Mathers of Mason City, Illinois, breeders of fine stock and remarkable men on many other counts. Both were Illinois University graduates, and Doc was also a product of the Chicago Veterinary College. There was so much to be learned from these two that Bill stayed with them for nine years.

At the end of that time he did some more thinking. "I was not a boy any more," he says, "and I thought it was time to start charging for my services."

Tom Dunn of Houston, who had a fine Shorthorn herd at Genoa, had been after Bill for a long time to come to Texas. Now he had bought up some brush land near San Antonio with the idea of building a good herd of Herefords. He persuaded Roberts to come down and manage the enterprise. They made good progress; but about the time they were ready to get out some publicity and start commercial operations, Dunn had a bit of hard luck with the government and found his ranch tied up with injunctions. The manager immediately became the former manager.

The next job looked like an ideal situation for a man who still wanted to learn. Just outside San Antonio, Tom Slick was carrying on all sorts of experimental work on the Essar Ranch. The most sensational of his ideas was the result of research in genetics which had convinced him that animal breeders were not putting their best individuals to maximum use. He had worked out a scheme for taking the embryos from his blooded stock, transferring them to grade cows for the actual production of calves, and keeping the superior animals at the business of producing more embryos—as many as six a year. Bill went to work on the Essar and learned many valuable lessons. But 1,875 acres and an oversized staff left him feeling that all his talents were not being put to use. He began looking around again.

149

By this time the top men of the cattle industry were becoming aware of Bill's capabilities. R. J. Kinzer, the venerable secretary of the American Hereford Association, took an interest in him and began questioning his former employers. Craig Logan, breeder of fine Shorthorns in Bosque County, Texas, heard R. J. ask Les Mathers one time if Bill was a good man.

"Any s.o.b. that works for me nine years," replied Les in cattleman's language, "has got to be *damn* good!"

About this time Charles Pettit, owner of the wonderful Flat Top Ranch in central Texas, found himself in need of a manager. O. R. Peterson and Jack Turner, both prominent in the industry, took Bill up to meet him. Kinzer was there, too. Bill and Mr. Pettit hit it off, and in October, 1941, Bill Roberts became head man of a really important cattle ranch.

Flat Top embraces 16,000 acres of range and farm land in some of the prettiest country in Texas. Nature did well enough when she raised the limestone hills and laid down the rich valleys in that region, but oil money and human persistence have improved considerably on what Nature accomplished. Charles Pettit has dammed the streams to make artificial lakes, protected the wild game, cleared off the brush, brought back the native grasses, and put the fertile patches into crops. He has built a palatial stone-and-tile ranch house, laid out eighty miles of improved roads, put up a sales barn in a picturesque location, and stocked the place with deep-bodied, broad-rumped Herefords. All of it is now under the supervision of the Canadian boy who started with one hundred dollars of borrowed money and a pair of unshined shoes.

Bill lives in a stone-and-frame house a mile from the headquarters with his Texas-born wife and his three children (two boys and a girl). He drives around in a red Mercury selling bulls, entertaining visitors, and trying to keep up with the thousand responsibilities of a ranch manager. It keeps him moving, but it is what he loves, and what he has learned to do. He is not through learning, either. At this moment he is breaking in a ranch veterinarian who has had training in animal dietetics. Bill wants to study animal feeding in all its aspects and arrive at some basic conclusions about chemical content and costs. Such an undertaking properly carried out, as Bill intends, may prove to be of service to the whole cattle industry.

"Do you hope some day to be an independent operator?" I asked him.

"Well," he said, with his gentle smile, "I am not without ambition, even now, but after working with the type of outfit I have become accustomed to, I—well, I'm satisfied to go on this way for a while longer."

The case of Bill Roberts brings up the subject of the ranch manager—a type of cattleman seldom heard of in the early days, except on the big corporation ranches. He is the product of the new era, a skilled operator in a highly specialized profession with a good deal of glitter about him. He has to know more, do more, and be more things to more people than almost any other human being connected with the cattle business.

Money with no place to go brought him into being. The years of prosperity since World War II have left quite a number of men with an extra million dollars or more. When one of these overloaded capitalists begins to wonder what to do with his hoard, he may decide that what he has always really wanted is a place in the country. If his hangover or his social activities are sufficiently burdensome, he goes shopping for a ranch in a remote corner of the range where there will be no telephones or cocktail parties, where he can loaf around in picturesque old clothes, and where the boys, unimpressed by the trappings of wealth, will call him by his first name. Usually he winds up with fifty thousand acres in Wyoming or Colorado or Oklahoma or Texas, acquires an expensive wardrobe of ranch clothes, rebuilds the run-down headquarters, and finds his chief recreation in looking at his expensive Herefords. It is really quite an experience to watch $25,000 of one's own money eating alfalfa.

Before long it begins to dawn upon him that beef is big business, just like oil or beans or railroads, and very shortly he is using the telephone he thought he wanted to get away from as he tries to make his investment (or his hobby) pay off.

Sometimes he makes it pay. Many amateurs with good luck and good help have made money and become convinced that they are cattlemen born. Others have found themselves in deep water. Help has been hard to get and keep; poor judgment and unskillful feeding have led to defeat in the show ring; wrong guesses about the market have made deep and painful incisions in the checkbook. When these things happen, an intelligent millionaire makes up his mind that he needs expert help and starts looking for a manager.

It still goes on, with no sign of a weakening demand. As long as oilmen, manufacturers, corporation lawyers, construction magnates, and movie actors persist in buying ranch property, there will be a constant demand for men to take charge. If and when the hobby ranchers and make-believe stockmen are forced to run for cover, we may return to the times when only ranch enterprises like the Waggoner Ranch or the Matador will be in the market for managerial talents. But those times are not yet.

Meanwhile, there are several ways of getting a manager's job. One is to go to college and specialize in range management. A quicker and surer method is to marry the owner's daughter. Some very successful managers have moved into the field by this route and more than held their own.

Another paved highway to a manager's job is followed by those who can arrange to be somebody's son and heir. Sam Lard, owner of the Ladder Ranch in southern New Mexico, is a Fort Worth businessman, but his son Sam has gone to Texas A. and M. College and is working into control of the ranch. C. M. Harvey of El Paso has put his son Phil in charge of his considerable ranch holdings.

In complete contrast to these ready-made managers are the small group of practical cowmen who have come up through the bunkhouse and the branding pen, equipped with nothing but cow sense and determination. One of them is managing the historic 6666 Ranch, 200,000 acres at the base of the Texas Panhandle erroneously supposed to have been won by Burk Burnett with a poker hand of four sixes.[1] In charge of the ranch, the

---

[1] Bob McFall of Wichita Falls, an old Burnett ranch hand, says that S. B. Burnett made a deal with a man on Denton Creek named Crowley in the early days. Crowley branded 6666 on the left side of his cattle and a 6 on the left shoulder of his horses and mules. Burnett was to pay a fixed sum per head for all the cattle he could gather. He rounded up about 125 and decided to keep the brand.

twenty-five hands, and the huge two-story stone house looking down on the little town of Guthrie is a friendly but harassed little man named George Humphreys (they leave off the "*h*" in Texas). His father was a cowman in Stonewall County who could give his son few advantages besides a herd of cattle to labor with. George quit school at fourteen and went to work for Lee Bivins and Bob Allen as a cowhand—moved on to a job with Gene Mayfield—and finally came to the Four Sixes.

He is a hard man to catch. The best way to make contact is to wait at the company store on the highway below the big house. Sooner or later he will be in, on the run, and will sit down for a little conversation in the office room at the back. For as much as ten minutes he will talk without absolutely going out of the room like a rocket, but it is best not to try to hold him longer than that.

Men like George Humphreys are being replaced by slicker types who can fit into a businessman's world as well as into the normal environment of the cattleman. Many of them are well educated, though not always for the job they are doing. Bob Anderson, employed by the Waggoner interests to manage the huge 3D Ranch near Vernon, Texas, graduated in law. Bob Lazear of the Wyoming Hereford Ranch, one of the best-known managers in the nation, finished an engineering course at the University of Michigan. In fact, he was thirty-one years old and working on an irrigation project at Grand Valley, Colorado, when he was offered the ranch job in 1920.

Lazear is a particularly interesting and unusual cattleman. His abysmal ignorance about livestock was offset at the beginning of his career by some previous connection with the WHR enterprise. His father had served for many years as a sort of super-secretary to the man who owned it—a Christian entrepreneur named Henry P. Crowell.

In his youth Mr. Crowell had arrived at death's door via a case of tuberculosis. The doctors told him he might live three years if he went to the West—three months if he didn't. He spent a few years in Colorado and Wyoming, and when he went back East again was so completely recovered that he lived to be almost ninety, dying in 1944. With him when he returned to the city went a brain full of ideas which made him many times a millionaire. Quaker Oats, Perfection stoves, and other enterprises were projects of his.

He always felt that his debt to the West was greater than he could ever repay, but he found a way to ease the pressure of the obligation. As a first step he bought the 57,000-acre WHR just outside Cheyenne. It was already stocked with good Herefords, descendants of five hundred animals imported from England in 1873. The second step was to ask Bob Lazear to take charge and start work on improving the breed.

How well Bob succeeded is shown by the fact that WHR blood is probably more widely disseminated than that of any other herd in existence. It has filtered out to the remotest corners of the United States, and to many another land as well, through the prize bulls and heifers that go to breeders every year. George Sessions Perry calculates that WHR bulls father thirty million pounds of beef every year *in the first generation*,[2] and there is no telling how many million pounds on the hoof owe their quality to the labors of Henry P. Crowell and Bob Lazear.

WHR animals have won acres of blue ribbons, tons of trophies, and miles of newspaper space. Nineteen times in twenty-five years the ranch has sent the Grand Champion Carload of bulls to the Denver show. Individual bulls have sold as high as $61,000, and Lazear won undying notoriety when he was given a guaranteed bid of $100,000 provided he would put his best young prospect, Helmsman III, up for sale at the 1945 Denver show.[3]

He turned the offer down. "I'm not certain what I could do with a hundred thousand dollars," he explained, "but I know what I can do out on the range with that bull."

Cattlemen understood what he meant, though few would have said "no" as firmly as he did. The worst thing a conscientious breeder can do, according to the unwritten code of the business, is to "sell his tops." To improve his herd, he must keep his best animals.

[2] "Bob Lazear and his Herefords," *Country Gentleman*, September, 1948.
[3] Lewis Nordyke, "Angel of the Range," *Saturday Evening Post*, November 3, 1945.

After such heroic action it was a crushing blow to WHR when Helmsman III lay down and died in 1947, the whole hundred thousand dollars' worth of him. But somehow his death made Bob's devotion to his ideals seem a little more absolute.

Purebred men have to learn to take their beatings in stride, and Bob has had his share. A large proportion of his troubles at the moment are due to the fad for small "comprest" type animals. WHR produces big-boned, amply proportioned stuff. The tide will swing back his way, Bob thinks, when the hobby ranchers and the bovine dilettantes begin to feel the pinch. "Cattlemen want size with quality," he believes. "These other fellows will get rolled when the business goes back into the hands of real cattlemen."

You will look a long time before you will find another enterprise run with the same object as WHR. In 1940, when Mr. Crowell began preparing for the end, he established a trust with three directors (of whom Bob is one) to administer the ranch and its affairs. He satisfied the Treasury Department that the entire income, less operating expenses, was to be devoted to the public good and got all the taxes remitted. Now every cent of clear profit goes into the charities and religious enterprises which Mr. Crowell designated before he died. Up to 1929 Bob was struggling to get out of the red, but since that time he has stayed on top and made a good annual contribution to the glory of God and the welfare of His people.

Manager Lazear is as unusual as the ranch he runs. He is a deeply religious man and takes an active part in church work. He does not drink, smoke, play poker, or cuss, though he never cramps the style of his cattleman friends who generally cherish at least one of those vices. Unlike the crusty cowman of tradition, he has a Sunday-school superintendent's ready geniality. One feels that it is a real geniality, not just an exercise in how to win friends and sell bulls. At the same time the tautness of a good business executive balances and restrains his amiability.

There is nothing of the empire builder about him. His frame is thin and wiry. His jaw is broad and square under a thin-lipped mouth. He dresses like the rest of us with no attempt to look like anybody special. When I saw him he had on khaki riding pants, high laced shoes of a vicious yellow color, a blue dress shirt, and a red bow tie. He works in an unpretentious room in the ranch office building, sitting behind an oak desk surrounded by filing cabinets, show trophies, and pictures of his cattle.

He leans back with one leg on the desk (he complains of a little arthritis lately) and talks about his three boys, one a Presbyterian missionary, one on the ranch, and the third in a ranch supply house in Des Moines. When he finishes, he will probably take you over the ranch in his 1937 Lincoln, charging up and down the steep hills as if he were driving an airplane. He loves to point out the irrigated hay meadows where a couple of thousand tons of winter feed are put up, the good bulls ready for sale, the ranch community of white houses and big red barns in the lee of one of those high-swelling green-to-the-top Wyoming hills. He is immensely proud of his little empire at WHR, and WHR seems to be immensely proud of him.

A special type of manager is required for the big company ranches which still survive—mostly in Texas. These outfits have been in existence for a long time, have weathered many a crisis, and have accumulated considerable tradition. Their object is to make money—not to serve as hobbies or contribute to the welfare of any members of the human race except the stockholders. An example of the kind of manager they look for is Johnny Stevens of the Matador Ranch, the southern half of the holdings of the venerable Matador Land and Cattle Company.

The Matadors are still incorporated in Dundee, Scotland, and have always operated on frugal, conservative Scottish principles. There is no nonsense in the minds of the directors about "employing modern methods," "keeping up with the latest research," or "conducting experiments." As long as the cattle are doing well, the ranch is not running down, and the enterprise is paying off as usual, nobody has an impulse to throw even the most modern monkey wrench into the machinery.

Murdo McKenzie, who in 1890 set the pattern approved by his Scottish stockholders, passed the tradition on to his son. John McKenzie now directs operations from a simple three-room suite of offices on the sixth floor of the United States National Bank Building in Denver. John, Jr., works with his father, along with Miss E. A. Hall, a sharp-tongued but warm-hearted spinster who has been with the firm thirty-five years, and Mr. Sam Wiley, a neat and courteous old Scot with a bit of Dundee left in his accent and a record of service as long as Miss Hall's. John McKenzie, a big man with a slightly curved nose and a precise, down-drooping mouth, comes to Texas about once a month.

Probably not many managers would have the peculiar combination of traits necessary for happiness and success under such a system as this, but Johnny Stevens seems to have worked out the right mixture.

Cattle were in his blood from the beginning. His father owned the Hash Knife Ranch between Throckmorton and Seymour, and consequently Johnny absorbed the intangibles of ranching as naturally as he learned to talk. When he was ready, he went to Texas A. and M., where he became a member of the International Stock Judging Team, but every summer he came back to work on the ranch. After graduation he went to work for the Pitchforks and stayed ten years, leaving when the manager who hired him got crosswise with the owners. "That's one thing I'll always be—loyal," he explains. "I didn't try to stay."

He found a berth on the Matador's north, or Alamocitos, ranch, and at the age of twenty-five became its manager. He was moved to Matador to run the 400,000-acre southern division in 1940, and there he still is.

You can tell by looking at Johnny that he is a driving human being, sparing himself no more than he spares his men and horses, and that is probably reason number one for his popularity with his Scottish employers. He gets things done. His own explanation is that he has learned to delegate authority. "When I was first hired as manager, I worked too hard," he says. "I tried to do everything myself. They told me they were not paying me to be a thirty-dollar-a-month cowboy. Now I train my men to do their jobs, and then let them do them. A good deal of my time goes into that sort of thing."

There is also the fact that, like John McKenzie, Johnny is a very careful man. Between them, not a sparrow falleth on the ranch that they don't know about. The sort of thing they have to watch out for came to Johnny's attention shortly after his arrival. On the Matador the chuck wagon is out almost all year, and is often twenty-five or thirty miles from town. "When I came," he remembers, "the cooks on the wagon were bad about running out of things, and everybody who came into town had orders to get something they were out of. I damn soon put a stop to that. I told them it was just as easy to think of what they wanted a week ahead as it was to bother everybody to bring things out every day. Now they are sometimes fixed for two weeks at a time and don't make any trouble for anybody."

Furthermore, he does not believe in the great storehouses of ranch supplies, including every bolt used in a windmill, of which some of the big neighboring ranches are very proud. "They have lots of stuff they will never use, that will spoil, or that will build up wasteful habits in the men," he believes. There is not even a company store or commissary on the place. Local storekeepers sell to the ranch at a discount, and Johnny does not think he would make any money by handling his own supplies. It goes almost without saying that nobody in the organization buys anything without a written order from him.

With this for a background, it will surprise no one to hear that the Matadors are conservative in their ranching methods; and Johnny Stevens, in spite of his college training, is as good a Scotchman in this respect as if he had been born in Dundee. Most of the cattle are handled with a horse and a rope just as they were in the old days. Johnny says it is a matter of time. Big calves may be run into a squeeze, but the little ones can be roped and dragged out much faster than they can be handled in a corral.

He does not have much truck with conservationists. Of course the management tries to keep grazing down and builds tanks to hold runoff water, but with over one hundred windmills on the place the tanks are not essential. Johnny will have nothing at all to do with mesquite eradication—doesn't believe in it —thinks the brush will come back thicker than before—and doesn't intend to have any clearing done until he knows for sure that other people have made it work.

As for accepting payments from government agencies for improving the range, neither Johnny nor his principals take

TEAMWORK ON A BIG RANCH
Bill Roberts, manager, and Charles Pettit, owner
Flat Top Ranch, Walnut Springs, Texas

much stock in it. "We might as well do the work ourselves," he thinks. "It all has to go back in income taxes if you don't take it off in expenses."

In short, Johnny Stevens is at home on the Matador because he believes in following the paths he knows and letting somebody else lose money by breaking new ones. Coy Drennan, foreman for the Pitchforks, puts it in words for both of them when he says: "I think the old-time cattleman's methods are best. It was by following them that he got to be an old-time cattleman."

Many a manager in the past has lost his popularity by watching his men and money too closely, but Johnny Stevens is one who can be cautious without arousing contempt. He is not a back-slapper—even seems dour and grim at times—but everybody in Matador seems to like him, and the greeting is "Hi, Johnny!" when he stops at a filling station or enters a hardware store.

Still young (he is thirty-five as this is written), he is a big, beefy fellow with almost-red hair (the kind they used to call "sandy"), and has oversized freckles on his hands, wrists, and the back of his neck. He is wide in the hips and getting a little thick in the waist, but his broad red face is still boyish, and so is the grin which infrequently breaks up the severe expression he habitually wears. He wears a gray shirt, riding breeches, forty-two-dollar boots, a Stetson hat, and a pair of gold-rimmed specs. He talks slow and sometimes waits a while in the middle of a sentence until he is sure his idea will come out the way he wants.

He could hardly be a more complete embodiment of the Dundee ideal unless he changed his name to MacStevens.

The career of a manager is not always a matter of moving onward and upward to the ultimate goal. Some men get in and out of managerships half a dozen times and sometimes take off in unexpected directions. Bill Ross, for instance, has gone through the cattle business like a broken-field runner on a football field, sidestepping, pivoting, and hugging the sidelines. At the moment he lives in the heart of the high cattle country of northern Colorado fourteen miles out of Steamboat Springs, but he was born in Aberdeenshire, Scotland, forty-seven years ago, and still has ball-bearing "r's" in his speech.

159

To find him, you take the road north out of Steamboat and wind higher and higher into the hills following the bed of the Elk River. The great Park Range stands off to the east, snowbound in April. Other mountain masses rise in chilly splendor to the northwest and southwest. The west-facing mountainsides are still patched with snow. Across the canyon, where the sun can work longer, the slopes are bare and wrapped in a soft-looking purplish garment of buck brush and service brush. As the elevation increases, the pines begin to appear, struggling up the cliffs in sturdy companies. Water soaks down through their roots and trickles across the road from every cut and canyon. All the bottoms are wet and every creek is brimming. At one place the highway crosses a raging little stream with a sign on the bridge reading "Dry Creek." Under the words somebody has written, "Like Hell." The Elk River, which crowds the road at every turn, is bank-full this time of year and charging downward like a herd of wild horses.

This is the turn of the season. In a month the bottom lands will be passionately green. The red wall of willows at the edge of the river will be green, too. The fields of oats will roll out a delicate green carpet, and the pastures, already turning, will be afloat with grass.

Outsiders are always spellbound when they see this country washed with springtime tints or ablaze with the brilliant reds and rich yellows of autumn. The natives are less enthusiastic. "It's an ideal country," they say, "if you like winter. We have nine months winter and three months late fall up here."

It suits Bill Ross perfectly. You find him happily at work in his barnyard suitably turned out in rubber overshoes, weather-beaten work pants, an ancient red plaid shirt, and a khaki cap with ear flaps. He has a square, honest face with a cleft in his chin and a blocky nose, the whole enveloped in a luxuriant underbrush of red whiskers. He has just had his upper front teeth pulled and is a little diffident about his speech, though he is an animated conversationalist and soon forgets about it. When he thinks of something interesting he beams all over and launches out with an explosive "By God . . . ."

The most interesting subject for him is cattle, and he comes by his interest honestly. His people in Scotland were all cattle-men and he grew up in the midst of the business. After World War I, in which he was too young to join, some of his friends came back with the feeling that Scotland was too small for them. At a local cattle show one day a restless former soldier said to him, "Bill, I don't think there is much place for us around here. I think we ought to emigrate."

Bill agreed. They flipped a coin to see whether it would be Australia or America. America won, and in a few short weeks the two were on their way.

Bill's first job was on a Shorthorn farm at Montevideo, Minnesota. It didn't turn out as well as he had hoped it would. Notes were changing hands rapidly and people were rich on paper. The man he worked for got caught, and so did Bill. He had been letting his wages accumulate, thinking he would study veterinary medicine at Ames, Iowa, but he never collected a penny. Fortunately a job turned up in California and he took the train west. He had just forty cents in his pocket when he arrived.

The months and years went by, and at twenty-five he was manager of the Blackhawk Ranch at Danville. By this time he was beginning to be known, but the road upward did not really begin to open until he started judging at some of the local stock shows. Eventually he worked up until he was officiating at events in the big cities, and at one of these—it was in Los Angeles—R. J. Kinzer showed up. R. J. was not merely secretary of the American Hereford Association; he was a minor deity in Hereford Heaven—a prophet and a seer. He liked Bill's work.

After the show he called the young Scot aside and asked him, "How would you like to judge our Ogden show?"

Bill was completely take aback. "I couldn't do it, Mr. Kinzer. I'm too young and inexperienced."

"Would you do it if I asked you to?"

There was a long pause, but finally he said, "It would be a great honor, and if you want me to, I will."

The Ogden show went off very well, and Kinzer came around again. This time it was, "Will you judge our Denver show?"

Outsiders can hardly realize what that invitation meant. It was like asking a local politician if he would mind stepping into the United States Senate without running for office. But by now Bill had gained a little assurance and he made Mr. Kinzer no trouble.

Since that time there has been no slackening in the demand for Bill's services. He is one of the top stock judges in the United States and has worked at all the big shows—Chicago, Denver, Fort Worth, San Francisco. His first call in 1949 was to judge the Pacific International at Portland, and the schedule from there on was heavy.

But it was never enough. Success, admiration, good money, and pleasant contacts all left him unsatisfied. He wanted a place of his own, and he wanted it worse as the years went by. Where to go was no problem. He noticed that the top animals of almost every show he judged were mountain raised. They had more hair and bone. The upland climate, he felt, was responsible; the cold winters developed vigor and hardihood, and the upland grasses probably gave more calcium and phosphorus. He said to his wife, a pleasant, dark-haired girl whom he had met in California: "We'll be in the cattle business for the rest of our lives, but we're in the wrong place. Let's go to the mountains."

She agreed, and has been well content up there in the green valley above Steamboat Springs, though she has had to give up many friends and advantages that meant much to her. As for Bill, he has never had any doubt that Heaven is located in the Colorado Rockies somewhere along the Elk River.

He found another managing job to tide him over while he looked around. John Barbey, maker of Vanity Fair women's apparel, had four thousand acres and 250 registered Angus cows in the Elk valley. The country was good and the cattle were good, but before long the owner and the manager had a difference of opinion about grass. Bill maintained that there had to be a balance between summer and winter pasture. He took the cows off the home place and trailed them up to the summer range about the first of June—a month later than Barbey approved. Most of the local stockmen were on Barbey's side, but Bill would not give ground. He had noticed that when the grass on the meadows was grazed a little longer it "stooled out" and made more hay instead of running to tall stalky tops.

Just below the Barbey place there was a small ranch owned by a widow named Mrs. Hallquist. "It fascinated me," Bill admits. "It was an ideal ranch setup, though it was all plowed up then and sowed to small grain." The widow moved to town and took to school teaching about the time the grass war started on the Barbey place, and Bill bought her out, twelve hundred deeded acres in all.

That was three years ago. Gradually since then Bill Ross has built up a herd of his own, sometimes buying fine stock that he has seen at the shows. He is proud of an eight-year-old bull whom he calls "The 44th," and of a few exceptional heifers. He would like to be able to fit about fifteen bulls a year for the Denver show, eventually. He intends to run his place on a hay-grain rotation and thinks the neighbors will have to come to this too. "Their meadows are getting sod bound," he thinks. "If the Forest Service makes any more cuts, which seems inevitable, these pastures will have to be handled differently, and that calls for fewer cattle, cross fencing, and better stock. A good grazer leaves as much as he takes off, and that means we ought to cull our herds while prices are high. When the pinch comes, we won't be caught with anything but our best."

With so many irons in the fire, Mr. and Mrs. Ross haven't got around to doing all they want to do with the new place. In fact, after two years they are still cleaning it up. The house is an old-fashioned two-story country dwelling with a lean-to in the back, like ten thousand others, only uglier because of the imitation-brick asphalt covering that has been nailed over the exterior walls. The Rosses plan to put up a new house across the road, but labor is hard to get and they haven't made a start.

They lead a pleasant, unhurried life. On Sundays they go to the Presbyterian church in Steamboat. Doug, the oldest boy, is a 4-H Club leader; the twins, Stuart and Linda, now eight, have two more years to go before they can join. There is a reasonable amount of visiting back and forth in the community, some card parties, some hunting over the hills.

Scotchmen and Englishmen have been winning favorable notice in American cattle circles since early times, and Bill is one who has done particularly well. "What is the reason," I ask him, "that you English and Scotch boys do so well in the stock business over here? How do you command so much respect from people who are not usually too fond of outsiders?"

That one costs him some thought. "Well," he finally ventures, "it might be because the boys from the old country are so faithful. In the old days one of them never thought of leaving his show stock. Always stayed right in the barn with them. If one had a calf, he was right there till she calved. You don't find that quality so much now."

Nevertheless, he has never been tempted to go back to Scotland. "After all," he says with deep conviction, "you know this is the best country in the world."

T 13. Dude Rancher

HE SIGN ON THE MAILBOX says "Broken Arrow Ranch," but from the looks of the symbol underneath[1] "Drooping Arrow Ranch" might be more appropriate. The main highway south out of Jackson, Wyoming, hugs the mountainside high above the Hoback Valley, and the ranch road plunges sharply aside into the depths, winds about looking for a bridge, finds one, and climbs strenuously to a bench on the other side. One gate to open, and there we are on one of the few real cattle spreads to which dudes are invited. The stout, genial, gray-haired matron in blue jeans on the front step of the main house is Mrs. Wallace Hiatt. Wallace will be off in the field somewhere soothing his soul with solitude and a piece of broken-down machinery.

It is really true that there are not many places left where, as on the Broken Arrow, a real cattle ranch and a dude enterprise flourish together. Every tourist camp in the West which rents out two broken-down horses to overnight guests calls itself a dude ranch; and if a couple of milk cows graze in the front yard, the proprietor has excuse enough to promise his guests some real cowboy experience. When the dude business started on Howard Eaton's horse and cow ranch in the Dakotas thirty years ago, the

[1]

promise might have been kept. Times were hard then, and many a rancher saved his hide by allowing tenderfeet, for a consideration, to ride with his hands. But more and more as time goes on, a dude ranch is becoming a specialized type of resort hotel, catering to people with ample bank accounts and inclined to be a bit snobbish.

"The rancher who has combined a stock outfit with a dude outfit," says Paul Van Cleve, president of the Dude Ranchers' Association, "is lucky."[2] He might also have added, "He is rare."

Real cattlemen are deeply suspicious of the claims of dude establishments to the title of cattle ranch. It is hard to get a real hand to work on one of them. Andy Anderson, of the A Bar A at Encampment, Wyoming, says ranch people ordinarily don't want jobs on a dude establishment. Rodeo cowboys and professional dude wranglers furnish the bulk of the skilled help.

The men in the business are well aware of this situation and try hard to give the real ranches the distinction they deserve. "There is as much difference between what we term a bona fide dude ranch and a roadside tourist camp as there is between a plumbing shop and a shoe store," says Walter G. Nye, executive secretary of the Dude Ranchers' Association.

The Broken Arrow is, to use Mr. Nye's phrase, "bona fide," but even so it doesn't seem a bit like the cattle ranches in the movies or the Western magazines. The main ranch house looks out over a lush valley right into the narrow gash of the Hoback Canyon which divides the towering Hoback and Gros Ventre ranges, and through which the Hoback River comes brawling and bouncing into the meadow lands below the house. On each side of the needle's eye of the canyon rise masses of snow-seamed granite with somber black cloaks of pine forest around their shoulders. Behind the house are gentler slopes clothed with a brighter green, but high enough to be snowy in June. It is not as raw and wild a mountainscape as the Grand Tetons farther north. A more endurable grandeur here takes the place of the brutality of those fierce igneous spires.

Mrs. Hiatt takes a practical view of her scenic attractions. "Our mountains are easier to ride into," she says.

[2] *The Dude Rancher,* January, 1949, 4.

The guest cabins slant off to the right of the main house, each having a full view of the valley and the Gros Ventres beyond. The Hiatts are proud of those cabins and say there are no prettier in the Jackson Hole country. They are built of varnished logs—double and single—with hot water, square bathtubs, dressing rooms, and closet space enough for the Duchess of Windsor.

The dudes who patronize the place are not the kind who turn up in flashier resorts. They have a family feeling about the ranch and about Mr. and Mrs. Hiatt. A year or two ago they were asked to express their sentiments about having a separate dining room and gathering place. They said no. They liked better to have their lounging and dining room in the main house. It made them feel more in the family. They had their way, and life on the Broken Arrow still centers in the big front room of the Hiatt's headquarters with a broad plate-glass window aimed directly at the narrow cleft of Hoback Canyon.

Mr. and Mrs. Hiatt have been dude ranching in the neighborhood for almost thirty years. At the beginning of his career Wallace was—of all things—a newspaperman, and a good one. He worked for Hearst in San Francisco and Oakland as an advertising man and editorial writer, went on to New York and Chicago, and finished by becoming bored with the whole business. In those centers of enlightenment he came to feel more and more that he was a "speck of dirt." Finally he broke loose and came west.

For two years he was a cowboy in Idaho south of Pocatello. Then he found the Hoback Valley and began in a small way in the dude business. On Friday, May 13, 1919, he started the Triangle F Ranch and store a few miles up the canyon from his present location. Little by little he built up his place until he was able to set up the Battle Mountain Ranch in the late thirties. In 1941 he began operating at the sign of the Broken Arrow.

He has a lot of land now, and has access to a great deal more. Behind him lies one of the great wilderness areas of the country. Sixteen miles out he keeps up a camp where parties go nearly every week for overnight stays. The place is high on a bluff in the midst of a primeval forest, and looks down on beaver ponds and some of the best fishing to be found anywhere. Hunting is wonderful too. Wallace and his men do not allow anyone but their guests to follow the ranch trails back into that country. The only barbarians who invade this sacred privacy are occasional long-winded strays who pack in from the settlements a hundred miles away on the other side.

There is always a serpent in Eden, of course. In this case the creature is the United States Government, specifically the Forest Service. The summer range on which Hiatt cattle run formerly carried fourteen hundred head. The carrying capacity has been reduced to two hundred—"not enough to make it interesting." To Wallace this seems rank foolishness, and he feels that his opinion is worth something because he really knows grasses. He even runs an experimental project on his own place in conjunction with the Soil Conservation Service and the Forest Service. The local ranger gives no heed to his protests, however, and when Wallace induces him to come out and look at the range he refuses to admit that there has been any mistake. Meanwhile the grass grows tall and constitutes a fire hazard which may, in the opinion of one former newspaperman, put the country back where it was when the Indians burned it off to get at the game many years ago.

The experimental plots are saving the day for him, however, and may yet do something for the whole grazing fraternity. He has plots of crested wheat grass, tall oat grass, smooth and mountain brome, Canadian and Russian wild rye, bulbous bluegrass, and a new Siberian variety called intermediate wheat grass.

This last, he thinks, is the wonder grass of all time. It grew five feet, six inches high last year by the middle of August, though the season was exceptionally dry, and was high in protein. By using this discovery, Wallace figures that he can bring his own range up to where he can handle his original herd without too much dependence on Forest Service lands.

Meanwhile the dude business goes on, but he leaves most of that to his lieutenants. First comes Shorty Kelly, the dude wrangler, a fine reliable cowboy and distinguished rodeo performer with just enough sense of drama to impress the guests without overwhelming them. He has a Texas drawl and a fund of salty idiom, is a good man with a steak over an open fire, and a top hand with horses. He cowboyed all over the West before he went into rodeo. After that he twisted broncs at the big shows from Madison Square Garden to Calgary. Now he works during the winters on the Rancho del Río at Tucson and puts in his summers on the Broken Arrow. His wife, Julie, is Mrs. Hiatt's chief assistant.

Next to Shorty comes Wally (Wallace E., Jr.) Hiatt, nineteen years old, 190 pounds, and all cowboy. He is also an expert mechanic and manages all the machinery.

One cowboy, Bill Archie, born and brought up on a ranch near Hartville, Wyoming, tends to the hogs, milk cows, chickens, and anything else that needs to be taken care of.

Wallace himself does the farming. He has 160 acres in one upland field and several other extensive patches where, he says, "I come out and dig in the dirt, which I love to do." The whole ranch is well mechanized, however, and neither Wallace nor anybody else absolutely breaks his back. "We try to keep it a sort of family operation," he explains, "so we have the machinery to enable us to handle it without much outside help. Power mowers, side-delivery rakes, a combine, and a hay baler are big helps. All the men have to do is load the bales in a truck and stack it in the hay lot."

They harvest two hundred to three hundred tons of hay per year and eighty to one hundred tons of grain—enough to feed their cows, horses, hogs, and chickens. "We butchered seventeen hogs last fall and got most of the meat left," says the proprietor.

It seems to be a good life for Wallace. In old hat and pants, ankle-length lumberjack boots and a fine red coat of tan, he asks nothing better than to get down under his plow and put on new points. He was fifty-seven years old last August, but doesn't look it. He is a lean Yankee type with ample nose and chin, and a wide, gap-toothed grin. He is a man of culture on one side and a man of the soil on the other (probably the happiest human combination), right spang in the middle of the tradition of Cato and Thomas Jefferson. He helped organize the Soil Conservation District and is all for progressive methods, but like all the rest of his tribe he despises bureaucracy and regimentation. He thinks the crowning abuse in his neighborhood is the spinelessness of the Forest Advisory Board. "We go on electing them year after year," he growls, "and it doesn't mean a thing. The local ranger can veto every suggestion the board makes."

In his voice one hears an echo of the cry: "No taxation without representation!"

Mrs. Hiatt is the one most involved in taking care of the guests and she likes to talk about her business. "If I had the gift of expression," she remarks regretfully, "I could write a book about it. The things I have seen and the characters I have met!"

"You mean the local citizens are more amusing than the dudes?"

"Oh, by all means."

She goes on to tell about such eccentrics as "Shoot-'em-up Bill," who lived up on the mountain somewhere, did no work, and made no money, but came to town two or three times a year and bought what he wanted with a fresh supply of twenty-dollar bills. He never came back with any change, either. She thinks he must have been a robber with a hidden hoard of stolen money, but Wallace didn't find any when he went up in his capacity of undersheriff to investigate the old man's death.

Shoot-'em-up used to tell marvelous stories, like the one about the time he went bear hunting. He saw a bear stick his head up over a log. He shot him, saw him fall, started to walk up to the log—and the bear popped his head up again. Bill knocked him over once more. This happened four times, and when he finally got up to the log, he had five bears.

"The whole community around Jackson is very special," Mrs. Hiatt declares. "You have to be there in the winter to see the population at its best. We have a couple of remittance men who do no particular work but are subsidized by people in the East who wish to see as little as possible of these knots on the family tree.

"There is still some of the old code in operation. When we came, it was not wise to talk very much. You might know what somebody else was doing, but sooner or later you might break a law or two yourself. One way to make a man get out his gun in a hurry was to ask him where that meat on the table came from.

169

It was likely to be the wrong kind at the wrong season, and a little too fresh.

"The temperature goes down to fifty below in Jackson. Donald Hough (who wrote *Snow above Town*) is right in saying that winter doesn't move up or down the country from Jackson. It moves straight up and stays there, though sometimes it hangs a little lower than usual. Natives can always feel how high or low it is. They can also tell when fall comes. They step outside one morning and sniff the air. 'Well, it's all over,' they say. Nobody else can tell it, but they know."

The Hiatts say that people come to their place to rest. They certainly do not come to flank calves, though the Broken Arrow is actually a cattle ranch. What brings some of them back year after year is probably the essential thing that a good dude ranch has to offer—a homely friendliness combined with graciousness and refinement. The atmosphere of the place is as far from snobbery on the one hand as it is from crudeness on the other. And anyone who had never been on a dude ranch would not expect it to be like that at all.

To show what a wide variety of attractions are covered by the term "dude ranch," take a quick look at the Grinning Skull Dude Ranch just outside of Dallas. In an old farmhouse decorated with wagon wheels and all sorts of skulls (including one of a rhinoceros) Ma and Slick Swope serve bountiful meals, entertain weekend guests, and run square dances.

"We bought a few flea-bitten horses—a few cows—all we could afford—come on out and started. We charge the boys and girls $7.00 from Sat. P.M. to Sun. P.M. that includes horse-back riding, ranch style supper—good bed, breakfast and lunch and a big dance on Sat. night where they meet other boys and girls and have fun. We have no drinking on the place and I think thats what made us"—so writes Ma Swope.

Ma's son Jimmy is musical and calls for square dances, which puts her ahead of the Hiatts in one respect. Shorty Kelly, their dude wrangler, has never learned to play the guitar, can't sing "Ridin' Old Paint," and never called a dance in his life. Fortunately the West is wide, and there is room for the Broken Arrow at one end of the range and the Grinning Skull at the other.

170

T

14. Cowboy Preacher

HE Reverend J. Stuart Pearce, D. D., Presbyterian minister, retired colonel in the Chaplain's Corps of the United States Army, and chaplain emeritus of the Old Time Trail Drivers' Association, is something we don't often see nowadays—a cowboy preacher. He was a puncher before he took up theology, and has gone back to the old cows now that his active career in the ministry is over.

What he calls his "thumbnail ranch" is just outside of Catarina, Texas, between the Nueces and the Río Grande on the edge of a farming community surrounded by an endless expanse of mesquite and prickly pear. His son handles a hundred acres of farmland, raising carrots with the help of artesian water. He himself takes care of 150 head of cattle on the rest of his 2,500 acres.

The first time I saw him, he was outrunning a strong-minded cow. A good rain had fallen the night before, and the whole country was swimming. I stood on the screened porch of his rambling white house and watched him running like a quarter horse and outmaneuvering the stubborn old beast he had just separated from her calf. When the cow was safely behind the pasture gate, he came back up the lane, a slender, wiry man in rubber boots who walked with short, quick steps like a small boy in a hurry—a small boy seventy years old.

He runs off all superfluous weight and has himself trimmed down to about 150 pounds, all bone, muscle, and hide. His high cheekbones, somewhat flattened nose, and wind-reddened skin give him an Indian look. When he gets his boots off and lies back in his chair with a cup of coffee, closes his eyes, and starts to remember, he could pass for a wise old chief if his knees were decorated with a blanket instead of a pair of his old officer's pinks. Gesturing with a pair of hard brown hands, laughing at the humorous side of his memories, bouncing up in his chair when he gets to the exciting parts, he throws a bridge of words backward across sixty years of time.

"No, I was not born here. I came to life near Pleasanton on the Atascosa Creek. My father joined a party and came out to

171

the Nueces east of Carrizo Springs when I was five. We were as poor as Job's old blue turkey and my father wanted more room to make a living for his family and try to get ahead.

"My people were all Scotch. Grandfather was a hardshell Baptist preacher who started out as a Presbyterian but apostatized and became a Baptist. I started out as a Baptist, too, but went over to the Presbyterians, correcting his error.

"It was pretty hard going. The Indians had been run out of Frio County and raided the country farther west until they were gradually all killed off. Nobody had any money. The men worked for the big outfits to make a little to tide them over the hard times and help them start again on their own when they went broke.

"My father went broke several times, but we were luckier than some. We always went back to the country south of San Antonio where my grandfather lived. Grandfather had a farm, and we worked on it until we could come back to the ranch, but we never admitted that we had been farming. That would have been a disgrace.

"We grew up with cows and horses and always thought of ourselves as cattlemen. So did everybody else, and they all lived according to the customs of the cattle country. My father never signed a note in his life. A man's word was all that was necessary. I never saw a front-door key till after I got married. We were poor, but we had as much as anybody else did. And everybody was welcome to what the rest had within reasonable limits.

"As kindly and hospitable as these people were, some of them had grown up in a pretty hard school. A number had come from farther east to put a stop to the feuds and killings that broke out after the Civil War. Joe Tumlinson and Doc White had played their parts in the Sutton-Taylor trouble. Alfred Allee had had his difficulties in the Frio country. His grandson is a captain in the Texas Rangers at Carrizo Springs right now. King Fisher's daughter lives there, too. Some of the Butlers came out, hoping for a peaceful life. They were all basically good people, too.

"We never did worry much about the minor vices. I can't remember when I first learned to smoke. My father had a paralyzed arm and couldn't roll his own cigarettes. The makings were kept on the mantel over the fireplace, and it was the job of the oldest boy to make cigarettes for him and keep them going as he crossed the room to where the old man sat. As each boy grew up and went away, the next in line took over the job. Pretty soon it came my turn. I remember one time hearing him say, 'Son, either you will have to roll two cigarettes or move faster. You have them half-smoked by the time you get here.' It was true too. I was smoking as fast as I could as I crossed the room.

"The country was mostly big ranches in those days. John Blocker had everything west and south, Asher Richardson and Old Man Votaw had everything east of him, and Tom Coleman had the rest. When I was fourteen, I went to work for Blocker. Many boys put in their summers punching for him. If they were good enough to stay on after September first, they were known as 'Blocker men'—and that was the best recommendation a cowboy could have.

"I was a pretty good cowboy. I could ride anything with hair on it and was a pretty fair roper, though I was at my best handling horses. Of course we all grew up on horseback and that stuff was pretty near second nature to us. I liked to read, though, and used to borrow books from Mrs. Tom Coleman. Her husband was one of the big owners. It was a hundred miles around his fence. I didn't realize it then, but she was getting in a little home missionary work. I read anything and everything and got a pretty good start on my education, though I wish now I had read a little more of the classics.

"There were cowboy preachers around then, mostly men of very little education. The one that had the most influence on me had never even been to high school. You might say the cattlemen were as short on religion as they were on education, and in one sense you would be right. They were not much on churchgoing, but they were reverent in their own way and sound in their moral ideas, no matter what they did or said.

"I found that out for myself in the fall of 1896 when I was about sixteen years old. In those days we shipped cattle from the nearest point on the railroad—Encinal or Eagle Pass—and John Blocker had taken a bunch of steers to Eagle Pass for shipment. After the cattle were loaded, the boss man would give the boys as much money as he wanted them to have—not all he owed them, because there would always be a bunch of chippy dens around, and they would come back drunk and broke. This time Blocker passed out the money and told the boys to be back at five in the morning, but he didn't give me any.

"When the rest were gone, he said to me, 'Jesse, you can have your money if you want it. You earned it and it's coming to you. But if I were you, I wouldn't go across the river.'

"'Thank you, sir, I didn't have any intention of going across.'

"'The boys won't expect you to. In fact, they'll be surprised if you do.'

"A lot of them were like that, and still are. When I want money now, I seldom bother my congregation. I go to some of my cattlemen friends. The substantial ones are the best backers the church has. The first gift of $25,000 I ever heard of was from Colonel Slaughter. So was the first fifty-thousand-dollar gift, and the first hundred-thousand-dollar gift. The first half-million I know of came from the Kokernots, and so did the first million.

"One of them helped me through school; I was the second boy in Dimmitt County to go to college. When I came back for my first vacation, Doc White stopped me on the street and put two twenty-dollar gold pieces and one ten-dollar gold piece into my hand. He said, 'Boy, you've got a long, hard row to hoe. You'll need help. Take this and pay it back if you want to. It's a gift.'

"I got through all right, majoring in systematic theology, and went to work as a Baptist preacher. My first charge was at Pearsall. That's where I met *her*."

Mrs. Pearce, a pleasant-faced, white-haired woman, has brought in a second cup of coffee by now. When he sees that she is listening, he lowers his voice and says behind his hand, "She made me marry her."

Mrs. Pearce has heard all this before and makes no objection beyond a faint clucking noise, but that is enough. "Let me alone," he shouts. "Get out or I'll divorce you in half an hour. If I didn't have the patience of Job, I wouldn't have stood it this long."

They swap grins with each other as she takes out the empty cup, and he says with a chuckle, "I had a friend with no sense of humor who used to tell me I ought not to talk that way—somebody might overhear me and think I meant it."

DUDE RANCH DINNER
At the Grinning Skull Ranch near Dallas
Jimmy, "Ma," and Slick Swope in the background

Photograph by William J. Davis

Story follows story, throwing light on a life experience that only a cowboy turned preacher could have had. There is one about a squatter (Jesse insists that no names be used) who got in the way of a big cattleman and was assigned to a second-rate gunman for elimination. The squatter turned out to be tougher than the pistol man, however: slapped his face in public, and made him back down. Jesse had just got home after a session at Baylor, and saw the whole thing.

Years later they both came to San Antonio when Jesse was about to dedicate a new church building and, for the sake of old times, they both attended church. As he sat on the rostrum that Sunday, Jesse saw to his horror that they had taken seats one behind the other. It was their first meeting since their run-in on the streets back home.

With mounting distress he watched the wife of the former squatter pass a note to her husband. He has the piece of paper now. It said, "—— is sitting right behind you."

The man hitched his belt around, and Jesse knew what he was carrying with him—he always carried it. But nothing happened during the service. Jesse turned the benediction over to an assisting pastor and had his plans made to dive through the potted palms in time to come between the men before anything could get started. He was too late, however. As the congregation rose, the squatter turned around and thrust out his hand. "Howdy, old-timer," he said heartily. "How are you?"

Jesse almost folded up on the front pew.

Nothing as exciting as that happened in El Paso, in Cisco, and in the other places where he preached and labored, though there were some violent ups and downs. Sinners, sermons, four children, and a tour of duty in the Chaplain's Corps during World War I filled his life to overflowing.

When World War II came along, he was rather old for active duty, but one of his military friends, now a lieutenant general, asked him to come back in. "He wanted somebody who knew what it was all about," says Jesse.

By the time it was over, his wife was ready to go into a nervous collapse, and when she was better, he felt like having one himself. He was ready to step out and into something a little

175

less strenuous, and began looking around for a spot to retire to in the country he knew so well. On the edge of Catarina he found a little ranch for sale with a big earth tank of clear water next to the corral behind the house. The tank sold him. Now he owns the place.

It is fun to go with him to Carrizo Springs where everybody is his friend. He waves at the biggest man in town and hollers, "Hi-ya, little boy!" Stahl and Smith, who run the Chrysler-Plymouth agency, were high-ranking officers in the last war. They address Colonel Pearce as "Sergeant," and he replies gravely, "How are you, Corporal?" He calls for a pair of boots at the local cobbler shop, sounds his horn for service, and embarks on a delighted argument when the grinning proprietor comes out and asks him what he thinks this is—a damn drive-in?

"I couldn't get along without my friends," he remarks as he drives off.

The standards and ways of the old days are still in him. He is a good lover and a powerful hater. And what he hates most is sham and fakery. People who write about the cattle country give him an acute pain. He tells about an Eastern literary man who approached him at a meeting of the Old Time Trail Drivers Association, wanting material for stories.

"Well," Jesse asked him, "if you want to write stories, why don't you go back to Philadelphia among the Damyankees where you know what it's all about? I'm getting tired of reading the stuff you people write and finding mistakes on every page, if not in every paragraph. The first thing a writer ought to do is master his background. I used to try to be nice to you fellows, but I don't think I will any more."

The writer said to another man later, "That's the roughest-talking preacher I ever saw."

When I was leaving, Jesse brought out some things he wanted me to see. The last was a pair of four-pointed silver spurs.

"I suppose every man has a few things he feels sentimental about," he explained quietly. "I feel a good deal of sentiment about these. They belonged to the best rider and roper this country ever knew. The old-timers all know what he was, though

176

the young ones never heard of him. Before he died, he asked his wife to send these spurs to me.

"I was always a good rider but only a fair roper. The only good man I ever beat in a roping match—and that was by accident—was the owner of these spurs. His name was Clay McGonigle."

Jesse Pearce personifies, as much as one man can, the religion of the cattleman, which is mostly a private and personal thing; usually straining at the bit to get away from organized piety.

In one Texas town it is said that the cattlemen sit on the curb outside the bank all week and go out to the ranch on Sundays so their wives can't make them go to church. Some of them say quite frankly that "Sunday is our day for riding and branding." Pious people point to the profanity, poker playing, and affinity for whiskey which are traditionally part of the cattleman's way of life, and assume that he is irreligious. The men themselves do not admit any such thing. "I figure your religion is inside, and it doesn't make any difference whether you go to church or not," one of them said to me. Jesse Pearce would understand what he meant, and would add that the cowman's Christianity is more dependent on works—on generosity, honesty, and neighborliness—than on faith.

The most successful efforts to get the stock grower into a church organization have been nondenominational. Ask Dr. George Nuckolls to tell what happened at Gunnison. In 1935 the Church Board of the Federated Methodist and Presbyterian movement decided to start a community church with no denominational backing. The idea caught on and the church grew. In 1938 it sponsored a Community Recreation Hall and Youth Center with a dance floor, lounge, pool tables, ping pong, and kitchen. Everything is free, and everybody is welcome. "It has worked out so well," says Dr. Nuckolls, a broad-faced, gray-haired, positive Dutchman, "that since the Youth Center went into operation we have averaged one trial in two years in juvenile court.

"I have never asked a soul for a cent. Every year when the people fill out their income-tax blanks they mail us a check. I have never sent out a notice or said a word about money in the

177

pulpit in fourteen years. I call on newcomers to the community, but I never ask them to come to church.

"The way I look at it, everything that helps life is religion. A lot of what we do here is preventive religion."

Those are principles that a cattleman can understand. He understands doing good. He understands the bigness of the universe and the insignificance of man. He likes to get together with his fellows once in a while for a religious service, and turns out by the hundreds for such events as the Bloys Camp Meeting which is held every fall in the Davis Mountains of Texas. But he distrusts glib piety and sentimental religious indulgence.

It is startling sometimes to watch him applying to such matters the principles he has learned in his business. For instance, there is the story of the new preacher who stood up before a congregation of three people to deliver his first sermon. The man was game. He would give them his best. So he preached his heart out. After the sermon he met his little flock at the door and asked one old cowman how he liked the service.

"It was all right."

"Well, I'm disappointed that you weren't better pleased. I really preached the best I knew how."

"Yes, I guess you did. But when I go out with the feed truck and only three critters show up, I don't pitch off the whole load."

A former cowpuncher would never have made that preacher's mistake.

T 15. Rich Man

HE MOST UNBELIEVABLE of all the big ranchmen is a fugitive from the Main Line in Philadelphia. Alfred M. Collins has managed the mighty Baca Grant, 250,000 acres of Colorado mountain and meadow in the shadow of the fourteen-thousand-foot peaks of the Sangre de Cristo, since 1930; but he hardly knew one end of a cow from the other during the first fifty years of his life.

His family was old, rich, and important. It was quite natural that he should be an officer in the First Troop of Philadelphia City Cavalry (George Washington's special regiment); that he should serve for years as captain of the Bryn Mawr polo team; and that he should become vice president and general manager of the family factory (paper specialties). Not quite so natural was the urge which made him a well-known explorer and big-game hunter.

He could write a book about that, and has actually traveled about a good deal lecturing on his experiences. He made his first expedition to Africa in 1898 and his last in 1928. In between, he explored in Asia, South America, and the Arctic, often heading expeditions sponsored by the Museum of Natural History, the Field Museum, or the Smithsonian Institution. He still keeps up his private collection of game heads and jungle treasures at the old home in Philadelphia. When he married Helen Glenn in 1928, he had the ceremony performed under the widespread ears and tusks of his prize bull elephant instead of before an altar, much to the dismay of his Front-Family relatives.

A relic of those times hangs on his living-room wall—a photograph of a younger Alfred Collins, intense, hawk-faced, a little lofty, and most elegant in white tie and tails, presenting the Elisha Kent Kane medal of the Philadelphia Geographical Society to Sir Hubert Wilkins.

It is a very special experience to sit with him among his rhinoceros feet, lion skins, elephant tusks, and books about far-away places and listen to the tales he has to tell. A concentrated, meticulous man, he sits with fingertips together and eyes far away, looking as if he were about to discuss the annual report with a board of directors. Instead, something like this comes out:

"He was the biggest gorilla I ever saw, and he came at me like a bullet. I hit him when he was only eleven feet away, and he kept right on coming. I thought I was gone when he struck me. The impact knocked me out momentarily, and when I came to, I hated to open my eyes. I was sure he was going to tear me apart. I looked around and saw him ten feet away, stone dead. My shot had killed him instantly, but the momentum of his charge had carried him right on over me. He weighed 450 pounds."

179

After World War I, from which he emerged a major, he found the family paper business disintegrating. He worried along with it for a while. Then one day he had a call and a proposition from the chairman of the board of directors of the Crestone Corporation, which controlled 100,000 acres of land in Colorado.

In 1823 a descendant of Álvar Núñez Cabeza de Vaca was given title to this tract—hence the name Baca Grant ("V" and "B" are interchangeable in old Spanish). It passed to Governor Gilpin, and from him to a group of Philadelphia capitalists, of whom Alfred Collins' father was one. There was gold in the Sangre de Cristos, but not as much as the company needed to keep such an enormous enterprise going. Eventually the directors decided to stock the place with cattle, and get rich more slowly and surely.

Again they were disappointed. The manager they hired had once owned a cow and had worked on the ranch for years, but he did not make the place pay. The directors were ready to start bankruptcy proceedings but decided as a last hope to call on Alfred Collins.

"I don't know anything about the cattle business," he told them. "Besides, I think this is a bigger job than you realize. I couldn't do it without giving up my own business, and I don't want to do that."

"Go out for the summer," they urged. "Take a vacation and make us a report on what you learn."

He accepted without enthusiasm, went to Colorado, and never came back.

He was fifty-four years old when he undertook to put the ranch back on its feet, and he did not know a thing about the business. Furthermore, the place was run down and neglected, the local ranchers cold-shouldered him, and for two years he had to put up with the old manager, who gave him the runaround whenever possible.

This, too, was at the beginning of the Depression, and he had a heart-breaking time floating loans, fighting the bureaucrats, and trying to get the ranch back in shape. Perhaps nobody but a Philadelphia industrialist could have pulled through.

All through these struggles Alfred Collins' watchword was

efficiency. "I didn't want to cut down the size of the herd," he says. "I wanted to raise twice as many cattle on the same acreage if I could."

Well, efficient production was what he had been trained for. It was just a question of applying the old ideas to a new field. He cleared off the brush, developed his artesian water supply until he had 150 flowing wells, learned surveying so he could lay out a system of irrigation ditches, cross-fenced, reseeded, rotated pastures, doubled and redoubled the productiveness of his land. He spaced corrals and camps five miles apart and connected them by telephone. He bought spraying equipment which eliminated the thirty-mile drives to the dipping vat. By putting calving cows in separate pastures and giving them hospital care, he increased his calf crop to over 90 per cent. "We believe that if a calf dies, it is our fault," he says. Finally he started keeping a good set of records, much to the disgust of the "practical" men who had been keeping it all in their heads, and building up a prize-winning herd of Herefords.

In 1945 he decided that he should cut down. He had been off the place for a short rest just once since 1930. So he held a dispersal sale of his registered herd—a sale which made cattle history[1]—and tried to get out of the purebred end of the business. Of course he couldn't give it up, and now at seventy-four he is up at five-thirty every morning, meets his foreman at six, and is all over the ranch with his men for the rest of the day. Nothing escapes him, and nothing amiss goes uncorrected.

As a businessman he hates any form of unthrift or untidiness. If he sees a tin can lying on the ground, he gets out of the pickup and removes it from the landscape.

Unthrift on a national scale bothers him even more. He thinks it is a shame that people with no brains and no initiative are the ones who count now. People like Henry Ford and Andrew Carnegie wouldn't have a chance. Herbert Hoover he regards as a great and misunderstood man. Our present economic situation he contemplates with pain and disgust.

All this makes Alfred Collins sound like an economic royalist

[1] In November, 1949, Noe's Baca Duke, sold by Collins in the dispersal sale of 1945, brought $65,000—highest price ever paid for a bull.

181

of the era of the Robber Barons, but it won't do to lump him with the Vanderbilts and the Harrimans. It is impossible not to like and respect this work-worn, high-strung, conscientious old man. He is an exacting boss, but a kindly one. He refuses to gold-plate his ranch. He wears riding breeches with a darned seat. He uses his organizing ability to help out any worthwhile project in his locality. He does not want to get rich. However, he thinks he ought to be allowed a little something for his efforts. Ninety per cent of the ranch income is gone when he has paid his income tax and corporation taxes.

"I don't want to hand my children a lot of money," he says, "but I'd like to leave them something."

His real object, he declares, is to build something and create employment for other people. If it is a sin to strive for efficiency, economy, and independence, then Alfred Collins is a very great sinner. But his cattleman friends seem to admire his brand of wickedness, for they approved completely when this product of the effete East was made Man of the Year in Livestock for 1949.[2]

E 16. Poor Man

VEN in these good times cattlemen are not all rich and prosperous. Next door to the big cattleman there is often a little one who wonders how, or if, he is going to make it. I met a man like that fourteen miles out of Rapid City when it looked as if I might be stuck for good in a South Dakota mudhole. . . . But this one needs a little build-up.

We turned off the highway, my son and I, four miles east of town and headed north across the lonely hills in the direction of Ernest Ham's place. It had been raining for a day and a night, and flurries of wet snow filtered down occasionally to plant moist little caresses on the windshield. The road was graveled and would have stood a stiff shower without turning a pebble, but these steady rains on well-traveled roads loosen things up. Wash-

[2] *The Record Stockman.* Annual Edition, 1949.

boards develop and deepen, slick patches of black soil appear, and a 1941 Packard behaves like an elderly hog on ice.

All might have gone well if we had been a little less literal minded. But Ernest had said nine miles to the turn, and when the speedometer said nine miles, we turned—up an ungraveled road with a heavy list to starboard. It was all over before you could say gumbo. Flirting her bustle like a venerable can-can dancer, the old car slid gaily into the ditch, found a hole where somebody had stuck before us, and settled solidly down on her axle. There was no way to get her out. We were thirteen miles from town, and the sky seemed to be puckering up for another good cry.

We could see Ernest's big white house a mile and a half away across the fields around an elbow in the road. A half-mile this side of the elbow was a smaller house which looked inhabited. Very low in our minds, we slopped and skidded in that direction.

Mud, weeds, and a few tufts of grass surrounded a small, square dwelling with a peaked roof. The place was twenty years overdue for a coat of paint. The roof of the front porch seemed to be looking for someone to fall on, and two floor planks were missing directly in front of the door. Around at the back was an outside cellar two steps from the kitchen. The roof had caved in and a heap of sad old junk was exposed below. It was an ideal trap for catching people or animals on a dark night, though no bodies were visible in the dark depths at that moment. Alongside the ruined cellar was a 1938 Ford Tudor with the left rear tire flat.

Vigorous breakfast-table conversation was going on inside, apparently among a large family, and at first nobody heard us knock. Finally a vocal obbligato by a big black dog, who had his doubts about us, helped us to become audible, and there was sudden and awful silence inside. A chair scraped back, and a youngish man in blue jeans and rubber overshoes appeared at the door. A pretty brunette woman in sweater and slacks looked over his shoulder and disappeared hastily when she saw strangers.

The man was definitely an attractive fellow, slender but well muscled, with black hair, good teeth, and a cheerful grin. He

183

said his name was Wayne Keith, and he was very sympathetic about our troubles.

"We'll get you out of there in a hurry," he promised. "We all get stuck once in a while in this country, and we know what to do about it. Wait till I get my coat."

It was not as simple as he thought. First we had to pump up that flat tire, and the pump was as leaky as a straw hat. We all got red faced and winded before the rim rose out of the mud. Then we tried to start the motor and found that the battery was dead.

"The middle cell is cracked," he said, "but it'll run if we can get it started. I'll have to pull it with the tractor."

He took off across the field in the direction of a battered old machine that crouched in a fence corner half a mile away. A nine-year-old boy, second of his five children, trailed after him, bundled in an oversize sheepskin coat. He had not said a word since he came out of the house, but stuck close to his daddy and watched everything he did. The tractor coughed a few times and finally started. Wayne and the boy rode it, roaring and bouncing, back into the yard. We attached it to the Ford with a log chain, and soon had the motor spitting and backfiring as we swung in a wide circle around the house.

While the rest of the rescue operation was going forward, Wayne told us about himself. He did not complain. He did not apologize. He did not make excuses. He just told us what had happened.

"I came over here from Deadwood," he said. "I grew up in the country but we never had much. When I got out of the service I wanted to start up for myself and found this place. They wouldn't give me a G. I. loan on it and it took all the money I had to buy it, so there wasn't anything left for improvements. We're really just getting started.

"We've had some hard luck, too. Yesterday my best heifer went out in the pasture and just laid down and died—I don't know yet what was the matter with her. She looked healthy enough when I turned her out, but when I went out after her she was lying there all stretched out. I just don't know . . . .

"We might make a little something on pigs this year, but the pigs haven't come yet. I couldn't get a boar when I needed one, and the pigging is late. We'll make it all right, but these things set us back pretty bad.

"I make a little working for the neighbors; always get along with them fine. Only trouble I ever had was with a fellow who wanted this place mighty bad and made it hard for me when I got it. He had some haystacks on my meadow when I moved in and he asked if I would let him leave them for a while. I told him I wasn't in any hurry—wasn't going to use the meadow right away. But spring came and those stacks were still there. I asked him to move them so I could turn my cows in, but he didn't do anything about it. Finally he took the top off of one of them and let it go at that. I told him to get them out of there or put in a fence, but he didn't do it, so I turned the cows in anyway. He moved them then, but he was pretty mad and tried to make me pay for the hay the cows had eaten.

"Then we had some trouble about fence posts. There were some steel posts in the meadow that didn't belong to me. I pulled them up and piled them in the yard and told him to come and get them. He complained when he came after them—said they weren't all there. Later on I found a roll of wire in the river with some posts on it, and I took and used them. I didn't think I owed him anything any more, but he was mad—said they were his.

"It came to a head in town. I was driving a pair of young horses that I was breaking. I was riding on the reach. He came out into the street and grabbed me by the collar. Called me a hard name and tried to hit me. The horses got excited and started to run—almost ran him down. He thought I did it on purpose and hollered at me, 'I'll get you, you ——!'

"I yelled back at him that any time he wanted to take it up when I wasn't driving skittish horses, I was ready. He could come by the house and we'd have it out; or drive to town. But he never did anything more about it.

"I hope I don't have any more trouble, because I want to make things go here. The place isn't very big, but there's lots to be done. I have 320 acres with 125 acres in small grain, 70 in hay, and the rest in pasture. It carries a cow to ten acres, about.

185

When I get on my feet, I want to have about sixteen dairy cows, eight or nine brood sows, and some beef cattle."

The log chain went around our axle and the Ford strained like a wrestler. In a shower of mud and gravel we came back on the highway. We drove our Ethopian vehicle behind him up to the house, and I asked him what I owed him.

"Not a cent!" he said. And no amount of urging would persuade him to let me do anything but make a birthday present to the silent boy in the sheepskin coat, who was nine that morning.

I asked the neighbors if this generous, cheerful, helpful fellow had a chance. "He is a fine worker and always ready when we need him," they said. "He may make it, but he has a long row to hoe."

PART IV It's a Hard Life

C 18. Trouble

OUNTRY PEOPLE depend on two things for their living: the sky and the earth. The one is unpredictable; the other is reluctant.

There is not too much a man can do about bad weather and difficult soil, except to take them as he finds them. Having learned this lesson, the ranchman usually goes along without fuss and lets Nature have her own way. If she says shiver, he puts on his old leather jacket and obliges. If she says sweat, he goes ahead and perspires without getting worked up about it. He is so used to this co-operation that such things as doors mean less to him than to other people. I walked into a store in the Texas cow town of Sheffield one time. The double doors were wide open on a frigid winter day, but a group of men were gathered about a hot fire in a pot-bellied stove inside. I wondered why they didn't close the place up. Well, they had a fire —why bother? They might as well have been squatting around a pile of glowing buffalo chips on the open prairie.

Men who were born in the ranching tradition seem to like to bring the outdoors indoors with them whenever they can, and try to eliminate as much of the indoorness as possible. This is sometimes hard on their wives, whose inside instincts are better developed. I knew one old ranchman in New Mexico who built his wife the new house she wanted but balked when she mentioned drapes for the windows. He would put up with no such frills and foolishness as that, and for years the poor woman felt miserable every time she looked at the naked sashes.

Living according to Nature's Simple Plan may cut down the overhead and make a ranchman less vulnerable to the upsets which go with his business, but no amount of co-operation with

189

Nature can make his life safe and easy. She has too many tricks up her sleeve. A man who survives as a cattle raiser for any length of time has endured all sorts of calamities and defeats. It used to be said that there never was a ranchman who didn't go broke. It might have been added that something threatens every day to break him. One year it will be loco weed or inkweed or some other poisonous plant increasing in his pastures for no particular reason. Then the wild game, protected by law, will tear up his haystacks. If he gets a good stand of grass just before a dry spell, he lies awake nights worrying about prairie fires.

He prepares for every emergency he can think of, but it is always the one he never dreamed of that defeats him. I know of one rancher in Arizona who fell victim to a plague of burros. When the mines in his neighborhood closed up, the owners simply turned their animals loose. They increased and multiplied, located the best grass, and soon took over his range. He applied to the authorities for permission to eliminate them, but the bureaucrats could find no regulation which authorized the slaughter. Finding that there was no other recourse, the burro-bedeviled rancher gave up and moved to Camp Verde, where no doubt he has found other troubles by this time to keep his mind exercised.

The Old Lady can cause terrific damage with the most insignificant instrument—for instance, the grasshopper. Unless one has been through it, he can hardly realize what a complete cleaning a man has suffered when he has been "grasshoppered out."

"In thirty-five, mȳ Godamighty, there was about twenty-two and a half grasshoppers for every foot of ground in western South Dakota," says Emmett Horgan. "There wasn't room enough for them all to sit down and they had to sit on each others' backs. They ate every green spear above ground and come mighty near getting down into the roots. We were grasshoppered out. But what do you think those government fellows did? I was out in the pickup one morning when I see a car parked on top of a little hill on my place. I rode over to see about it and there was a man on the ground with a square yard all staked out. I asked him what he was doing, and he said, 'Counting the grass.' I said to him, 'You know damn well there's no

THEY RUN THE BACA GRANT
Boots Muller, Lewis Vanderpool, Vogel Sandelin,
Rufus Denton, Bill Hutchinson, Perry Miles, Leon
Carpenter, Jim Burke, and Alfred M. Collins

WATER GRASS, AND BEEF
Cattle on the Baca Grant, Crestone, Colorado

grass here. The grasshoppers ate all of it. There's not a spear left.' And he said to me, 'If it wasn't for you goddam cattlemen there'd be plenty of grass.' Can you imagine!"

A cow is a moderately tough animal. A goat, they say, is always looking for a hole to crawl through, and a sheep is always looking for a place to lie down and die. Old Bossy is not as undependable as that, but she has to be watched constantly, for her brain power is low and she has a genius for blundering into trouble. She bogs down in mudholes, tangles herself up in barbed wire, or gets sick from eating something she should have had sense enough to let alone. If the soil contains insufficient mineral, she will gnaw on tin cans and bones and get them stuck in her mouth. Then she has to be run down, roped, and taken care of. "We had one old steer," says Henry Blackwell, "that was looking pretty drawn and gant. I sent the boys out to look for him when the cattle came out of the brush in the evening, but he wasn't there, so I went to look myself. Finally I saw his head just in the edge of the brush, and when I came up to him I saw he was too weak to move. He had a shoulder bone wedged in his cheeks and hadn't been able to eat or drink for days."

A whole chapter could be written on rattlesnake bites and their consequences to horses, cattle, and men. A boy who gets mixed up with a rattler seventy miles from town is in a bad way. It happened on the Harral ranch on the Pecos one time when a Mexican boy was struck on the leg. They called the doctor on the telephone, put on a tourniquet, and raced for town in the pickup. Just after they reached the highway, the fan belt broke. They flagged down a car and kept on. The boy cried when they tried to loosen the tourniquet, and circulation had been shut off for an hour and a half when they finally got him to the hospital. To make matters worse, there was only one dose of serum in town, and they had to send to El Paso for more. Eight injections of El Paso serum kept the victim alive, but he limps to this day, and Mrs. Harral never has got over the experience.

These minor emergencies ebb and flow within the larger patterns of disturbance—drought, dust, flood, hail, and economic depression. The bad times have always come and always will come—sometimes to one part of the range, and sometimes to another. Every few years a dry spell or a bad winter; terrible loss of stock; grass grazed back to the roots; rolling clouds of dust; bankruptcy! The wary look in a cattleman's eye is compounded of equal parts lore of the cattle business, memories of personal difficulty, and forebodings for the future. He knows he can stand hardship, and he knows he is going to have to.

On the plains the great spectre has always been drought. Every eight or ten years there has been a bad one in the north or in the south, or both.[1] In the nineties—the twenties—the thirties, the rain refused to fall, year after year. The cattle died like flies. Prices went down to nothing as steers with no place else to go glutted the market. Stockmen lost everything, or gave up, or died.

The memories of those times are bitter yet, and rarely does a man like Goob Saunders come up with anything good to say about them. "I came into Killdeer one time," he reminisces, "with a size six shoe on one foot and a size eight on the other and fifteen cents in my pocket. That was in 1933–34, the worst time we ever had in Dakota. The government was buying stock. I stayed in town of a couple of days shipping what I had, and that was when I found out the value of friends. I ate well and slept in a bed every night, and when I left town I spent that fifteen cents on a can of Prince Albert tobacco."

After the drought comes the dust. On the southern plains in the thirties the whole country turned to gray powder which drifted and blew like snow. When the first dust storms came on around Liberal, Kansas, people thought the Day of Judgment had come. Bernard Limmert tells how his fences were covered with sand, his chicken coops were buried, his doors blocked so that he had to shovel to get in or out.

The heartbreak and suffering of that time can still put a wild look into the eyes of people who tell about it. Jimmy Keating has a story about Dan Raines (that is not his real name) who hit bottom and almost stayed there. "Dan was a laborer," he says, "and a pretty good old boy. But he got so hungry he broke

---

[1] *Conservation on the Western Range*, U.S.D.A. Regional Information Service, *WR Leaflet No. 103*, 1: "Weather records show an average of 1 to 4 years of drought out of every ten years."

into a warehouse in Liberal and stole some of that government fruit. They caught him and put him in jail. He never denied taking it—just said he was hungry. I went to the authorities; told them Dan was all right—that he was hungry and couldn't be blamed too much. After he got out of jail, he came out to the ranch and asked me for a job.

"I said, 'Dan, I can't give you anything. There isn't any work here, and if there was I wouldn't have any money to pay you.'

" 'But I'm hungry!'

" 'I know you are, Dan. I can see that without you telling me.' Dan's ribs were actually sticking out.

"Mother went out and killed two of our chickens and fed him. He ate them both. He worked for me a couple of years with no more wages than his board and chewing tobacco."

If anything is worse than drought and dust, it is bitter cold and deep snow, and those come along periodically too. The great storms of eighty-six and eighty-seven wiped out three-quarters of the livestock on the northern ranges, say the old-timers,[2] and half a dozen calamitous winters have brought death and destruction since. Each time it has seemed like the end of the world to the men who were hit hardest.

At the meeting of the Montana Stockgrowers Association in 1887, President Joseph Scott rose and addressed the pitiful handful of men who had come in to pool their miseries: "I am glad to see this attendance, notwithstanding the fact that walking is bad. I am proud that we have such an attendance; I think it is proof to us that we are not to bury the large industry as some have stated, but we are here to revive it, and we are here to see that it does not die. It is true, the chilling winds of last winter have been felt on the range and in many places you can smell the dead carcasses in the canyons; but the case is not as bad as it might have been."[3]

The old men kept quiet about the winter of 1887 after the big snow of 1949.

[2] "Storm Stories Recall Early North Dakota Blizzards," Bismarck *Tribune*, February 10, 1949.

[3] E. A. Phillips, MS account of the founding of the Montana Stockgrowers Association.

On Sunday morning, the second of January, newspaperman Charley Thompson got up at six o'clock as usual in his big yellow house on the corner of Twenty-fifth and Evans in Cheyenne, Wyoming. He looked out of his bedroom window and saw a vast quantity of nothing—swirling white nothing. The houses across the street were completely blotted out by the first great onslaught of the January snow. He did not see those houses for four days. When the visibility returned to normal, the alley diagonally across from his front door was buried under sixteen feet of snowfall. Somewhere underneath was his neighbor's car, nobody quite knew where. On the Wyoming Hereford Ranch east of town the snow was seventeen feet deep between the office and the barn.

The storm struck Chadron, Nebraska, at 5:30 A. M. and went on to blanket the Nebraska Panhandle, northern Colorado, and all of Wyoming with the worst assault of killing weather since the eighties.

It was totally unexpected. Saturday had been chilly, but the temperature had held above freezing in most places. The weather prediction for western Nebraska was "Cloudy with light snow."[4] Instead of a light touch of winter, however, everything north of the Arctic Circle seemed to descend on the range country at once. At Chadron thirty inches of snow fell in thirty hours. The weather men admitted that they had been caught with their predictions down.

If some of the Indian population had been consulted, the shock might not have been so complete. Billy Circle Eagle, who ranches on the Cheyenne River Reservation in South Dakota and sometimes writes pieces for the papers, reminded the editor of the Sturgis *Tribune* of this fact later on:

"All Indians well know this month February call the name of this moon Lookout Moon this moon always have bad blizzard Mr. Indian always know day before was blizzard come. They know by signs first see the northern light in the night they know differend kind of weather signs. Sundog in morning storm come that night Sundog shooes in the moon storm come before morning."

4 Omaha *World-Herald*, Blizzard of '49 Edition, January, 1949.

Trouble

The blow was so sudden and severe that no warning or preparation could have made very much difference. For three days the snow fell heavily while the thermometer plunged far below zero and sixty-mile-an-hour winds whipped and slashed the country.

It was suicide to be out in it. Philip Roman, his wife, and his two children were frozen to death in northern Colorado as they tried to follow a fence to the safety of a neighbor's house. There was a break in the fence through which they wandered into a field where the creeping death overtook them. Mr. and Mrs. Andrew Archuleta and their five-year-old daughter died in their car near Hillsboro, Wyoming, and when the rescue party reached them, the car was completely filled with snow.[5] Sheepherders froze to death beside the road. Old folks died of fright and exposure. According to newspaper reports, six people perished in Colorado, eight in Wyoming, and two in western Nebraska. The loss of life among livestock and wild game was fearful, and as people looked out of their windows at the twenty-foot drifts and the solid wall of driving snow, they wondered if anything could possibly survive out in the open.

All communication and travel came to a sudden halt. Half a dozen transcontinental trains were caught in the drifts and passengers were marooned for as long as four days. Travelers on the highways took refuge in country stores, filling stations, and ranch houses to put up with short rations and wait for the snowplows. The snow stopped on the afternoon of January 4, but after a one-day respite the wind charged in again, roaring and booming and piling the drifts higher. On the eighth, new storm warnings were issued. On the tenth, another blinding terror struck Utah and Nevada.

Meanwhile the citizens had been doing their best to meet the situation. Planes belonging to the National Guard, the CAA, and the private operators took to the air on Wednesday to locate stranded travelers, watch for distress signals from isolated farmers and ranchers, and carry the sick to town. Rescue crews toiled off across the wastes of snow.

County Agent B. H. Trierweiler couldn't get out of his house

[5] Cheyenne *Wyoming Eagle*, January 7, 1949.

195

in Torrington, Wyoming, Sunday or Monday. On Tuesday he managed to get to his office, plunged into rescue work, and for five weeks afterward hardly ever got home. Planes went up over eastern Wyoming on the sixth. Radio station KOLT at Scotts Bluff, Nebraska, broadcast messages which were relayed to ranchers with no telephones. A set of signals was worked out to be used by those who needed help. The planes dropped food and cans filled with kerosene to people in need. Overnight the whole country was full of good neighbors.

By January 12 the suffering of the animal population began to crowd in ahead of the human problems. As new snow came on the fifteenth, blocked roads and drifted fields were making it impossible for the feed trucks to get through. Ranchers who owned bulldozers opened up trails to their haystacks, only to see them blown full by the constant blasts of the terrible wind. On main highways the drifts were packed so hard that they had to be dynamited.

All the way down the backbone of the continent the same icy paralysis had set in. On the sixteenth, thirty cars were reported stalled in the San Augustine Pass in southern New Mexico, and farther west on Highway 80 as many as two thousand travelers shivered it out in their cars until they could be pushed, pulled, or hauled to town after hours of terror. Lordsburg was swamped with several hundred victims who bedded down in the Knights of Pythias Hall and in the local churches. Nothing like this had ever happened in New Mexico before, and New Mexico was not half as hard hit as the northern plains. At Ogallala, Nebraska, over two thousand travelers—almost one for every resident—fumed and stamped their frostbitten feet and overflowed the town.

The great wind blew with hardly a lull. It blew on the seventeenth, the eighteenth, the nineteenth, the twentieth. County commissioners met in prolonged sessions and could do nothing. On the twenty-first, Governor Val Peterson of Nebraska met with the legislature to report that it was costing the Highway Department $200,000 a week to fight the snow and that he would have to ask for half a million for relief in the distressed areas. Next day he sent out a call for heavy snow-moving equipment to be rounded up wherever it could be found. Bulldozers, cater-pillar tractors, and trucks came in from five hundred miles and more away to join the fight.

Snow was falling fast in South Dakota that same day when Governor George T. Mickelson appealed to the Fifth Army Headquarters in Chicago for help in breaking out the roads to avert starvation for people and livestock in his state. Every-where food and fuel supplies were running low, and there was no way to bring any in.

That night the battering wind came up again, filling the new-ly cleared roads with snow and the bulldozer men with despair. On the twenty-sixth, Governor Peterson telephoned President Truman and asked that Major General Lewis A. Pick be placed in charge of emergency operations. The President responded by asking Major General Philip B. Fleming, Federal Works ad-ministrator, to handle the emergency program, but Fleming had a speech to make in Florida and flew off to keep his appointment. He was photographed in his shirtsleeves soaking up the Florida sunshine, to the complete disgust of the hard-pressed snow fight-ers. It was not until January 28 that Pick was given the word to go ahead, and not until the twenty-ninth that the President au-thorized the expenditure of the necessary funds. As General Pick set up his organization in Omaha, the headlines were screaming: "Worst Blast of Winter Grips Texas." The plains and mountain country from Montana to the Río Grande had almost reached the end of the rope.

General Pick, builder of the Ledo Road and at that time Mis-souri Division engineer, had been expecting the assignment and had already made some preliminary surveys. He called in Ad-jutant General Guy N. Henninger of Nebraska, who had had the National Guard officially in the field since the twenty-sixth, and Operation Snowbound got under way.[6] With local citizens as guides, courageous pilots to run interference, bulldozers to clear the roads and break out the haystacks, weasels to carry supplies across the snowdrifts, and trucks to follow up, the great army of snow battlers swept westward.

They reached Torrington on February 12 and found that

6 Lincoln *Sunday Journal and Star*, Blizzard of 1949 Edition, January, 1949.

the work of rescue and supply was already organized and going full blast. The Relief Commission was meeting in almost continuous session and a convoy system was bringing help to ranchers who could not help themselves. It was all worked out on a blackboard and set up on a master sheet. Clearing stations had been designated where convoy drivers could pick up orders for coal or oil or feed. The county had bought a new snowplow and a heavy truck. At 6:00 A. M. on Monday, January 29, thirty-four vehicles had assembled to form the first convoy, and had struck out for Jay Em at 6:32. The National Guard had tackled the job of clearing the roads. There was not much left for the army to do.

In western Wyoming the Reclamation Service came in early and did heroic service. Heavy machinery was moved down to Rawlins from projects at Riverton and Lander in charge of five drivers who worked, in some cases, as much as twenty-four hours without rest. Melvin Lynch, the county agent, went out with them as a guide, got lost with them, and suffered with them through the darkness and cold. They worked at such jobs as digging out thirty-two cars stranded between Rawlins and Casper, bulldozing a road to bring three thousand sheep to the railroad, and taking supplies to dozens of snowbound country families. The army came to their assistance late in the emergency period.

Meanwhile truckers were hauling in hay from far south on the plains, and the railroads were unloading feed along the right of way, wherever the cattle or the feed trucks could be brought to the tracks.

Most of it was paid for. The Red Cross provided some emergency supplies free of charge; contractors lent machinery, and individual ranchers gave food and shelter to truck drivers and weasel operators; but Operation Snowbound was not a charity bazaar. The army asked nothing for its services, of course, but it gained some very valuable experience without having to arrange maneuvers. General Pick reported after his force was liquidated that he had plowed out 87,073 miles of road, opened up communications for 152,196 people, and provided access to feed for 3,598,638 head of livestock.[7]

[7] Omaha *World-Herald,* Snowbound Section, February, 1949.

The army went at its job with so much dash and got so much publicity that the work of local agencies and the heroic efforts of the plain people were practically overlooked by the rest of the country. Spectacular gestures like Operation Haylift, which dumped thousands of bales of hay on the backs of astonished cows in blockaded pastures, were played up in spite of the fact that the effectiveness of such measures was rather doubtful. Ranchers in areas where the haylift functioned say that it was as hard for the stock to get at the baled hay as to reach the buried haystacks. They feel that it was not the Air Force but the man with a bulldozer and a truck who brought relief to the hungry heifers.

"We did not sit in the house and worry," says Jim Grieve of the Dumb Bell Ranch near Independence Rock. "We got out and fed every day, with the feed truck following up right behind the 'dozer and the road drifting full behind us. When we were out in it, it seemed fantastic. A man could hardly believe it was really that bad—thirty to forty below for three weeks and the wind roaring night and day. How anything lived we still don't know. Another ten days of it and nothing would have been left alive."

It was rough going, but experienced ranchers were never in as much difficulty as the papers made it appear. Archie Sanford at Alcova had plenty of supplies, though he ran a little short of cattle feed and butane. He fed his cows from horsedrawn hayracks and stayed on the place quite comfortably until a neighbor came over with a bulldozer and opened up the road out to the highway.

The men whose fathers used to go to town twice a year for supplies were ready for anything, as usual. They think that anybody who is stupid enough to get caught without a backlog is either a fool or a greenhorn, or both. Down at Chadron, Hud Mead tells about one young couple who ran out of food on the second day. "They were used to buying ten pounds of sugar and twenty-five pounds of flour at a time—they would even drive into town for a loaf of bread," he says.

"Yes," adds Mrs. Mead, "and that's the reason they don't have anything. They spend all their time driving back and forth."

The ones who helped themselves resented, and still resent, the way all the credit was given to the army. The National Guard, the Reclamation Service, the Soil Conservation men, the employees of the Department of Agriculture, the veterinarians, doctors, Red Cross workers, nurses, county officials, and plain honest people in every part of the stricken area were doing everything they could all the time, and most of them are firmly convinced that they could have made it without outside help.

Russell Thorp asked the chairmen of the local relief boards to prepare statements for filing in the office of the Wyoming Stock Growers Association, and most of these men were less than enthusiastic about Operation Snowbound. One county agent reported that "relations were strained" because the army seemed to feel that the local people were trying to get something for nothing. He recommended that in future "emergencies be taken care of locally."

Nobody who lived through those days of trial minimizes the debt the country owes to General Pick and his men. The objectors are simply reacting like old Charley Collins who once entertained the pastor of his church at his Colorado ranch. Charley was an extra-good cusser, and his wife warned him not to shock the Reverend if he could help it, so he kept a tight rein on his manners as he showed the visitor over his place. "The Lord has certainly blessed you," the preacher kept repeating as each improvement was pointed out. "The Lord has done wonders for you."

Finally the old man couldn't stand it any longer. "Well, by God, old Charley Collins has been pretty busy out here for the last fifty years, too!" he exploded.

Some of the stories in circulation show that a little selfishness and meanness did crop out. It took eighteen hours to plow out one family near Rawlins and get them to town. The husband returned to his cattle, but the wife and daughters stayed in and went to the movies. After a couple of days, however, urban life began to pall and they decided to go back to the ranch. They called the courthouse and said that the food situation on their place was getting desperate. The bulldozers cleared the way for them again, but one sharp-eyed driver noted that the only supplies they took back with them were some bags of popcorn and cookies.

Another woman sent word that one of her sheepherders was down with an infection and would lose his eye if he didn't get to town. They went out and got him, found that he was suffering from nothing worse than a cinder, and on investigating discovered that the woman merely wanted him off the place so she could hire another herder. They made her pay for the trip.

Everybody had his story to tell, and some of those tales will be retold by many a fireside many a year from now. People kept themselves alive by burning their fence posts and their furniture. They ran out of food and lived on field corn for days. They crawled out of second-story windows when the houses were completely covered with snow.

Bob Lazear says that the wind was so strong it blew the hair off his bulls. They turned their tails to the wind, and the icy particles driven through the air at fifty or sixty miles an hour cut at the hair until the bottoms of some of the animals were practically bare.

Wonderful anecdotes were exchanged of cows trapped in snowdrifts for as much as twenty-six days without food or water who came out under their own power and went to the herd; and there were sadder stories of herds which died to the last cow. Near Ashby, Nebraska, 150 head wandered out on the glassy surface of a lake. One by one they skidded and fell, and not a single animal was able to get up again. In spite of such gruesome episodes the total loss for the whole period, according to most estimates, was under 5 per cent.

The great storm had its humorous side, too, and some of the best stories have a chuckle in them. Melvin Lynch tells how he went out with a National Guard weasel one day when the snow was falling thick and fast, the visibility was zero, and the wind was blowing about seventy-five miles an hour. One of the drivers insisted on coming in spite of a badly upset stomach which gave him frequent trouble. He was wearing voluminous army flight pants which had to come down every time he got out of the weasel. When he got back in, those pants would be blown full of snow right down to the ankles, and his companions had to go to work and clear them out with a shovel before he could pull them back up again.

Another man whose story was told in several Nebraska papers woke up during the night of the big blizzard when he heard a disturbance in his henhouse. He stepped into his shoes, seized his shotgun, and strode forth into the storm protected only by his long underwear. At the door of the chicken house he leveled his gun and got ready for action just about the time a fierce gust of wind came along, caught the back flap of his drawers, and opened them up like a barn door. His old dog, who had also come looking for prowlers, moved in from the rear at that moment. As a result the man and his wife spent the rest of the night dressing the thirteen chickens that he accounted for with one shot.

One thing sure—many people learned a hard lesson from that winter. Wilford Scott ran out of butane during the first bad days while he had two children down with chicken pox. He now has a Skelgas tank in his yard nearly as big as a boxcar. Another Nebraska rancher spent two months trying to get off his premises, and when he finally reached town he bought ninety dollars' worth of groceries.

It seems that a ranchman's hardships have their educational value.

D 18. Rustlers

ICKENS is a pint-sized cow town thrown together casually in the midst of the immense pastures at the base of the Texas Panhandle. The biggest thing in it is the courthouse, built like a cracker box and not much larger, where the county's 8,500 people bring their grievances, sins, and taxes for final reckoning. The county officials do their business in leisurely West Texas fashion on the first floor. The second story is mostly given over to the courtroom—the theater where the last act of many a rangeland drama has been played.

The place was crowded on February 29, 1949, when Delbert Bailey took the witness stand to tell how he had helped to steal fifty yearlings from John Guitar, on whose ranch he was foreman. The crowd was there partly because big rustling cases are not common any more, partly because Delbert had been around a long time and had many friends, and partly because they didn't care much for his boss. Guitar was an absentee owner who lived in Abilene, did not keep up his fences, and had consistently underpaid his foreman ever since Delbert started work on the ranch in 1934 at thirty dollars a month with a wife and four children to feed.

As Mrs. Lucille Allred, the court reporter, looked around the room, she could feel that the spectators were on Delbert's side. She noted, as he took his place, that he was a six-foot cowboy with no hips to speak of and a pair of wide shoulders under a red plaid shirt; that he was scared to death; and that there was something wrong with his neck. His chin rested in a cup attached to a rigid frame which kept him from turning his head. Everybody else in the courtroom knew that he had broken his neck in a fall from a Guitar windmill. It was also known that he had broken arms, legs, ribs, and teeth in line of duty.

Three others were charged along with him—his brother-in-law, Bob Dillashaw of Lubbock; his helper, L. D. Bond; and a Mexican boy of respectable connections in the neighborhood named Raleigh García. According to the testimony, Bond was the villain of the piece—a small, dark, determined fellow who fought the case tooth and nail and did his best to pin the responsibility on the others.

Bond was working in the corrals one day with Bailey and García when the idea hit him. They had been late branding that spring and on account of worms and bad weather had quit before all the stock was handled.

"I bet we could steal fifty head of these calves and never get caught," Bond remarked meditatively. "They'd never be missed."

Bailey turned the idea over in his mind, which did not work very fast. "It would be risky business," he finally said.

"Well," Bond went on, his decision already made, "I'm going to get some of them. If it's all right with you, Raleigh and I will haul them off and sell them. You don't have to know anything about it."

Delbert didn't say yes. On the other hand, he didn't say no. "If I get caught, I'll take the blame," Bond promised. "We'll split the money three ways."

"It's up to you," Delbert told him. "If you get caught, I'll lay if off on you."

The first load went off on May 3 to a sales ring at Lubbock, five head in all. Delbert got less than fifty dollars out of that load. This, with about seventy dollars that came back to him from a later haul, was all the pottage he received for selling his cattleman's birthright.

A few at a time they slipped the calves out, some to Lubbock, some to grass pastures farther south. Bond got a number that Delbert never knew about, but most of the time the foreman was a willing helper.

In June they took Dillashaw in with them. Delbert told about it in his "Voluntary Statement":

> During the rodeo here in Spur last June my brother-in-law, Bob Dillashaw, who lives at Lubbock was down here with a load of beer, and he was at my house at the ranch. He and L. D. Bond and I were talking about making off with the calves there from the ranch, and Bob asked me to let him in on a cow deal, and we told him we might do it. Some time after the rodeo Bob Dillashaw rented a trailer in Lubbock, and came down to the ranch one night after dark. He came to the house and woke me up and asked me if I was ready to load that yearling. He knew that I was figuring on him coming. I got up, helped him load a calf I had put in the lot. . . . Our trade was that he would sell the calf and split the money with me, but I have never got any money from him for this calf.

In his own statement, Dillashaw told how he had got behind in his rent and had persuaded his innocent landlord to take half of the beef from this yearling to apply on what he owed. He hauled three loads in all, the last one bringing him around $225. Delbert did not get any of that, either.

Fifty missing calves make quite a hole in a herd of 800, and John Guitar could not be kept in ignorance. He reported his loss, and the Cattleman's Association took it up. They had poor luck. The last person anybody thought of connecting with the depredations was good old Delbert Bailey. Sheriff Kimmel, however, refused to be outdone. At his urging the Association men kept on checking and asking questions, and finally they made up their minds that it was an inside job. The rest was easy. Confronted with the evidence they had scraped together, Bailey broke down at once and told the whole story.

He was pitifully eager to tell all on the witness stand, but had trouble playing the game as the lawyers wanted him to. He sweat and stammered under the unaccustomed mental strain as he tried to understand what they were asking him, and they had to put their questions into words of one syllable. Delbert had cow sense, but he needed more than that on the witness stand.

On the second day the jury brought in a verdict of guilty. Dillashaw and García each received a three-year probationary sentence. Delbert thankfully accepted a probationary five-year term. Bond got two years in the penitentiary, where he is now putting in his time.

What happened on the Guitar Ranch and in the Dickens courtroom is fairly typical of modern rustling and its consequences. Ask a sheriff or brand inspector anywhere in the West if much stealing is going on in his territory and he will say no— the officers have the situation well in hand. Press him a little further, and he will admit that he has had a few reports of missing stock, "but we are not absolutely sure the stuff was stolen." Stay with him, and he may remember a special case a few years ago; and after that he will let his hair all the way down and go into detail about the nights he has camped out in the pastures, followed trucks that went nowhere, found the hides he was looking for burned up in a vat of acid, and otherwise pursued the elusive rustler with less than moderate success.

The fact is that stealing, most of it the work of small-time operators, is going on all over the range country at this moment, but it is hard to catch the thieves—harder still to pin anything on them or send them to jail. Cattle theft is still a thorn in the cowman's flesh, and will continue to be as long as cows are worth money and not too hard to steal.

It is not as bad as it used to be, of course. The day of the rustler dangling at a rope's end is gone—forever, we hope. The mavericker disappeared with the open range. Stockmen no longer make a point of eating somebody else's beef. "Winchester men" hired to protect the herds "and no questions asked" have gone the way of the big roundups. No cattleman now would think of paying a drifter to "brand strays" as a means of raiding his neighbors' herds.

One Texas cattleman advertised a few years ago that he would pay $250 for the conviction of anyone caught stealing his stock. An old cowboy was heard to remark when he heard of the offer: "Not so long ago they were paying me five dollars a head to steal them. Now they are offering $250 to catch me."

But though the stealing dropped off after the vigilante days of the eighties and nineties, it never stopped and has shown repeated signs of becoming a real menace again. In the late thirties and early forties it caused the worst kind of worry to cattlemen's organizations, sheriffs, and legislators. "Cattle Rustling on the Increase!" the headlines shouted. "Cattle Rustling Can Be Stopped!" other headlines shouted back. It was not brought under control, however, until some spectacular cases had aroused much public interest. The secretaries of the stockgrowers' associations in the cattle-producing states can tell blood-curdling stories of those times and have the records to go with them. Not everything about some of these episodes can or should be told, but a few samples can hardly fail to give a better understanding of the cattleman and his problems.

Fred C. DeBerard, a top Hereford breeder at Kremmling, Colorado, was in trouble in 1937. Five of his expensive cows had disappeared without trace. He reported the loss to Dr. B. F. Davis at the Stockgrowers and Feeders Association office in Denver, and a Pinkerton man was sent out to look around.

It was supposed, and correctly, that the cows had been killed so that somebody could get possession of their calves—a terrible waste since the mothers were worth at least a thousand dollars apiece and the calves would bring maybe a hundred. The number-one suspect had some outstanding little Herefords in his herd, too good to go with the rest of his stock. But to prove his guilt was another matter.

In his progress through the neighborhood the detective ques-

4-H CLUBBER AT WORK
J. D. Jordan of Art, Texas, grooms his calf, while
Willard, Lois, and Ethel May look on

tioned one old man who was very sympathetic. He said it was terrible about this killing of registered stock. He had lost some himself. Later the operative found that he had been talking to the father of the thief.

Still later he took another trail and stopped at a ranch house to pass the time of day. The owner seemed to be a pretty good sort, and the Pinkerton man asked if he had any suggestions about the matter in hand.

"If you had gone as far in the other direction, you'd be close," the man said. That was the first real clue.

The next clue turned up in Big Timber Gulch. A coyote had scratched up some bones in a gravel pit. The detective dug a little deeper and found the remains of three cows. One horn was branded 523, a number belonging to DeBerard. At Davis's suggestion the gravel near by was sifted, and three bullets were found.

That made a case. The suspect was arrested, bound over, and scheduled for trial. The day before he was to appear in court, however, he was coming down a hill with a truckload of coal which overturned and killed him. It was probably a lucky break for him.

The following year the newspapers got hold of something that appeared to have even more of the old juice in it. They had heard about a sort of Rustlers' Roost out on the lonesome plains east of Walsenburg where thousands of hides and other offal had been dumped into a deep and secret canyon.

"Hundreds of rotting cowhides found on the floor of an isolated canyon, disclosed to brand inspectors and police officers Monday the fate of many cattle missing from Southern Colorado roundups in recent years"—thus reported the Denver *Post* for September 13, 1938.

"The smell of the rotting hides was vile, and officers climbed to the bottom of the Canyon as frightened buzzards noisily took to the air. The isolated country is inhabited only by bobcats, white tail and mule deer, antelope, owls and eagles," said the Pueblo *Star-Journal* on September 11.

"The officers declared that the now bloody canyon must have been the secret rendezvous of a rustling ring that has been operating for years," added the excited reporters, and wondered if the thieves might not have been getting the stuff out by airplane.

The investigation began rather quietly. J. A. Busch of Rattlesnake post office stopped a truck on September 1 to see what was in it. Busch was just a small rancher and had lost only a few cattle, but apparently he felt his loss more than the big ranchers who surrounded him felt theirs. The driver was an Italian meat dealer in Walsenburg, and his truck contained the carcasses of three beeves. He said he had bought them from a near-by cattle raiser.

Busch accompanied him into Walsenburg and turned him over to Inspector Young Farr, who had the rancher brought in. The man told a good story. He said the meat was from his own cattle—that he had butchered them and sold the hides to a Pueblo man whom he described minutely to the last whisker. The officers went with him to Pueblo to look for this mythical character, but though they used up a lot of shoe leather, they never found him.

Meanwhile, several inspectors and peace officers had gone out to look over the ranch and had come across the canyon which set the newspapers to hunting for lurid adjectives. Their report showed that the stories had been greatly exaggerated. The reeking canyon was just an ordinary slot in the rock. The mountains of hides shrank about 1,000 per cent. It was true, however, that some butchering had been going on and that shredded hides had been laid on ledges in the canyon or covered with rocks and dirt.

The best evidence found by Inspector Harry Beirne was twelve severed legs which he matched with the fore and hind quarters of the carcasses in Walsenburg.

The rancher, who was an old Texas cowboy and a fine one, got out on bail and tried to work up a defense, but it did him no good. He spent the winter in a cozy cell.

Western Farm Life paid its respects to the rustling situation on March 1, 1940. It told of a rancher from South Park in Colorado who came home from a trip to Denver to find the head of one of his steers in the middle of the road outside his gate—

right where he couldn't overlook it. Twenty of his cattle were missing. "These human wolves have become so brazen that they enjoy taunting their victims," said Editor Cross.

So much stealing was going on in Colorado and other states that something drastic had to be done, and something was done with the passage of the McCarran Act in the fall of 1941. Transporting stolen stock across a state line became a federal offense and brought the F. B. I. down on the necks of the rustlers. Large-scale stealing became a very risky business.

A big case was waiting in Colorado to put the new regulations to the test. In December, Manager H. D. Mitchell of the 246,000-acre Trinchera Ranch in the San Luis Valley found himself short 109 head of cattle. He reported to the State Board of Livestock Inspection Commissioners, and men were assigned to the case at once.

The first break came when A. E. Headlee of Monte Vista noticed that twelve head of stock that he had recently purchased resembled pictures of stolen Trinchera cows. He reported this fact and also the fact that the cattle had been sold to him by two brothers who lived across the fence from the Trinchera pastures. Brand Inspector Harry Beirne soon learned that the brothers had been shipping cattle in trucks to Kansas to pasture on wheat belonging to Curtis Weaver of Manter. Beirne went to Kansas early in January, and on his return to Denver called in the F. B. I. The men were arrested and placed in the Alamosa jail.

A good many people were distressed about it. The father of the suspects was a pioneer rancher, seventy years old at the time, who had once been worth half a million dollars. He and his people had always been regarded as substantial citizens. But something was wrong with the boys' management, and they began losing out. Among other financial difficulties there was the matter of a $4,500 loan—payments due—and no money. When some Trinchera strays came on the place, they were not returned, and others were rounded up and put under the fence to join them. Some came from former Governor Billy Adams's ranch twenty-five miles away.

The fifty-one head of Trinchera cattle found in Kansas displayed brand alterations which, said the commentators, "were

the work of an artist." The Trinchera brand was a down T slash.[1]
The brand blotters used a wagon tongue.[2] By imposing the
wagon tongue on the down T slash,[3] a very convincing combination was made. The brothers were practicing on it up to the very
day of their arrest.

The men pleaded guilty at their trial during the first week
in March. The younger one was sentenced to three and one-half
years, and the older to five years in prison.

The McCarran Act was a help, but it was no cure-all. It could
not eliminate neighborhood stealing, sporadic butchering, and
occasional misbranding, all of which still go merrily on. It is so
simple and easy to pull up beside the fence of a remote pasture,
shoot a yearling through the head, throw him into the pickup,
and be two hundred miles away by dawn. This happens hundreds of times every year, and it takes the hardest kind of detective work and usually a lucky break or two before the officers
come anywhere near the thief.

A claim for reward, filed by Mrs. Harry Dayton with the
Wyoming Stock Growers Association, shows how a crook is sometimes caught. When F. L. Duvall found one of his milk cows
butchered on April 20, 1941, Sheriff George J. Carroll began
questioning the neighbors to see if anyone had noticed a car entering or leaving the pasture. Mrs. Dayton had seen a car and
described it minutely, even to the stickers on the windshield
and rear window. Carroll immediately became sticker conscious
and after a week's search spotted the car in Cheyenne. The owner tried to explain it away, but finally owned up and was sent to
prison for three years. Without those stickers he might never
have been located.

This sort of beef hunting is always most prevalent where
some sort of transient population—oil workers, road crews, construction gangs—has invaded the country. Earl Stewart, undersheriff at Meeker on the Colorado Western Slope, thinks that
much of the stock stolen in his neighborhood turns up in butcher
shops in the oil-boom town of Rangely. At least fifty head disappeared in the winter of 1948–49, but all the officers had to

210

show for months of patrolling was one dead porcupine, the victim of a wild shot when they heard a suspicious rustling in the bushes.

Sportsmen ostensibly after deer sometimes turn out to be cow hunters, and ranchmen have learned to watch these Nimrods closely. "Every year Montana livestock producers are taken for a rough ride by the soft-cushion rustlers who are clever enough to trespass on your range in their automobiles with or without your knowledge, load a chicken or calf or lamb into their car trunks and then thumb their noses at any law enforcement officer who tries to search their automobile," says one indignant writer. "There is a law on the state's statute books which could prevent this if only enough public opinion could be exerted to give the individual law enforcement officers the support they need to search cars and make arrests when this law is being broken."[4]

Royal Brinkerhoff tells about one beef hunter who rode into his camp in the mountains east of Bicknell, Utah, probably expecting to find nobody at home. In some confusion he asked about a man whom he was obviously not looking for, and rode off. "I took his track," says Brinkerhoff, "and after a while I saw him and another man fishing. I rode by just so they could see me. They pulled out and went farther upstream. I followed them and rode by again, and they left the mountain. When a man goes after beef, he knows he is doing wrong and is half-spooked already. If he knows somebody is around, he goes a lot farther before killing a cow. The real hunters, the ones in red shirts and caps, don't do so much harm. They scare our cattle half to death and it's months before we can get near them, but they don't do it on purpose."

Any peace officer or lawyer in the cattle country can tell a dozen different stories about beef hunters who came to grief or escaped punishment. One of the best ones, as told by a Southwestern judge, concerns a commissary sergeant at a military post. This man got an allowance of several hundred dollars a

---

[4] John Willard, "Soft Cushion Rustlers," *The Montana Stockman*, November 15, 1949, 8.

month for meat. He used to eat his steak and have it, too, by slipping through the fence into a neighboring pasture and butchering a beef occasionally. Somebody began to suspect him of irregular conduct; the inspectors were called; and the captain in charge took them to look over the meat locker. What they found there they called fresh steer meat (they can tell cow from steer by a certain muscle in the loin). The captain told the sergeant to wrap up that piece of meat and wait for him in the truck. They took the evidence to a cold storage plant and impounded it there. The soldiers at the base all chipped in a dollar apiece, $300 in all, and retained an attorney to defend the sergeant.

The case came to trial. The sergeant kept telling his attorney to make them produce the evidence—make them produce the evidence.

"You're crazy," the attorney told him. "The evidence is there. You're just running your neck into the penitentiary."

"It's my neck," the sergeant replied.

"All right, have it your way."

The lawyer thereupon began needling the jury.

"A lot of people say they've seen the evidence, but it hasn't been produced in court," he reminded them.

Finally they produced it. The inspectors swore to the string and the wrappings—they had not been tampered with. But when the package was unwrapped, there on the meat was the purple stamp of a legitimate dealer.

The sergeant was turned loose on this evidence, to join the great and glorious company of those who have beaten the rap. And there is still conjecture as to whether or how he substituted packing-house cuts for the fresh meat when he wrapped it.

Not all the stolen cows are headed for the meat market. A more intimate kind of stealing is going on wherein neighbors, and even relatives, attempt informally to share the wealth. Mrs. Dixie Stewart and Mrs. Elsie Haverty, who work for the Livestock Sanitary Board in Phoenix, Arizona, get a good laugh out of one important rancher who happened to drop in at a sale ring and spotted several of his own steers which had gone under the hammer. He immediately laid claim to them and hollered loud and long that they were stolen. The purchaser just as indignantly declared that he had bought them, paid for them, and had the papers to prove it.

"All right," roared the rancher, who was practically frothing at the mouth, "you get those papers and let me look at them!"

His name was on the bill of sale, all right, signed by his son who lived on another ranch. The father cooled off faster than a Texas cowboy in a blue norther.

Johnny Stevens of the Matador tells about one of his neighbors who ushered a few cows through his fence, changed the V brand to an M, and appropriated the calves as they came along, never making any attempt to sell the mothers.

Other examples of old-style brand blotting come to light from time to time. A recent case involved the changing of a lazy E[5] to a window sash.[6] In 1942, when there was considerable excitement over rustling along the Gulf Coast of Texas, a more imaginative brand burner was arrested near Mercedes. He had accumulated a number of cows branded with the bow and arrow[7] and by a simple twist of the iron had changed the impression to a "flower pot."[8]

In his Cheyenne office, Russell Thorp used to display an extraordinary relic of the blotter's art. In 1939, J. L. Bainbridge of Bristow, Nebraska, sent part of his stock off to pasture on a ranch run by two boys and their mother. At the end of the season they came up with a short count and said the missing cows had died. They could not produce the carcasses, and Bainbridge told Thorp about it.

The case was cleared up when Brand Inspector Earl W. Carpenter became suspicious of a peculiar-looking brand at the Denver yards. He called headquarters and Thorp told him to have the animals killed. Bainbridge's flag brand[9] showed plainly on the inside of the hide. The old lady back on the ranch had conceived the idea of rounding the corners of the flag, adding two letters and producing the 7PC. She was also smart enough to sell out and leave the country when she learned that the cattle were being held back at the yards.

In spite of the vigilance of owners and officers the problem

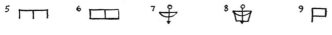

of control is not getting easier. Cattlemen themselves are to blame for some of the difficulties. They are often careless about proper branding, and some of them who should be called farmers rather than ranchers do not brand at all. In addition, they fuss and fume when the regulations cause them any inconvenience. The Gill brothers, for instance, have recently added the old Dana ranch on the Wyoming-Montana line to their holdings in Texas and New Mexico. The law says that stock cannot be moved from one state to another without inspection, and makes no exception for outfits with pastures on both sides of the state line. Consequently when their foreman moved two milk cows from Wyoming into Montana, he was arrested and the Gill brothers were quite disturbed. They have hired lawyers to try to do something about that law, though in the long run it is one of their most effective protections.

Ranchmen sometimes make strange mistakes, too. They come rushing into Association offices and report losses which may or may not turn out to be bona fide. Chase Feagins, chief brand inspector for Nebraska, went out to check on one herd which was reported as being thirty head short. Instead of looking for tracks, he asked for a count. It came out two head long. He noted that income-tax reports were about due at the time.

Of course, only one man in a thousand would try such a trick. Ordinarily the man who reports a loss has really lost something. His trouble is not made lighter, either, by the fact that convictions in rustling cases are always hard to get. Juries hate to send friends and neighbors to the penitentiary, and politically minded judges hate even worse to hand down unpopular decisions. In Colorado, says Dr. B. F. Davis, the district attorney used to have the sole right to recommend probationary and suspended sentences, and his recommendation usually stuck. Recently this right has been turned over to the judge, and several cases have shown that some Colorado judges have limber knees.

In Utah, adds Association President L. C. Montgomery, the situation has been very bad in this respect. In one county alone not very long ago eight rustling cases came up before the District Court and not a man went to jail. The cattlemen met for their annual session shortly afterward and agreed to try some new tactics. At the next term of court they gathered and packed the courtroom. They never said a word—just sat there and looked at the judge and jury. "We stiffened them up," says Montgomery, "and after that we had some convictions."

Lee Evans of Marquez, New Mexico, ran into another difficulty. Lee ranches a long way from town in the midst of a predominantly Mexican population. "I live out where they still carry the mail on horseback," he explains. In 1946 he was rapidly being stolen into bankruptcy. He knew where his cattle were going, however, and when he reached the end of his patience, he took a spade and a few friends and began digging around a certain house. He turned up a great quantity of hides, heads, and hoofs, and identified thirty-four of his own cows, eleven of them carrying unborn calves.

The defense attorney stressed the fact that the accused was a poor Mexican and Lee was a "rich" American. The jury, 100 per cent Mexican, turned the suspect loose.

"A Mexican may have worked for you for thirty years," Lee says, "but he still won't convict another Mexican. They think you're rich and they don't mind seeing you lose."

For seventy years the cattlemen have been doing everything they could think of to discourage and punish cattle theft, and in most states where stock raising is a major industry they have succeeded moderately well. Kansas and Oklahoma need to have their brand laws overhauled. Nebraska and South Dakota have recently given theirs a much-needed stiffening. Texas, Colorado, Wyoming, and Montana are proud of the way they have worked their problems out. In all these states, and in the states farther west, brands must be registered, and the brands must be inspected when the cattle change hands. A state board of some kind is charged with the responsibility—the Livestock Sanitary Board in Arizona, the State Board of Brand Inspection Commissioners in Colorado, the Brand Board in South Dakota—but in every case the state organization of cattlemen is and always has been the engine which powers the machine. This cattleman's union offers rewards, hires detectives, trains inspectors, co-operates with peace officers, keeps records, and sees that nobody goes to sleep. In Wyoming, the secretary of the Stock Growers

Association is also chief brand inspector. In South Dakota, the Brand Board, through a sort of hidden-ball play, appoints the Association as its agent, and the Association arranges for a chief inspector and a staff of assistants. This makes Emmett Horgan, the president, directly responsible for the efficiency of the program.

Texas has the most elaborate, extensive, and independent setup, with most of the work resting on the shoulders of Henry Bell, secretary of the Texas and Southwestern Cattle Raisers Association, and his men. The Association employs seventy inspectors. Forty-three are designated as market inspectors and are located in fifteen leading markets to which Southwestern cattle are shipped. They make records of brands, locate cases of doubtful ownership, and in general protect the interests of members of the Association, who, incidentally, do not all live in Texas.

From March 1, 1948, to February 28, 1949, market inspectors handled 2,709,227 Texas cattle. Of this number 1,182 head were reported out of line somehow—not all stolen. Proceeds from the sale of 200 of them were collected and turned over to the real owners. Restored to the real owners were 283 more. The total sum saved for members was computed as $139,166.12.

During that same period twenty-nine cases involving cattle theft were tried with evidence supplied by Association men. Twelve resulted in penitentiary sentences of two to five years each. There were four suspended sentences and five probationary sentences, six acquittals, and two hung juries.

One would think these statistics would give the cattle thieves pause, but apparently rustlers are not statistically minded. The little thefts continue, and even big stealing is not entirely a thing of the past. Roy Braidfoot of Deming, New Mexico, had a whole truckload of cattle stolen in April, 1949. The thieves were caught at San Angelo, Texas, and twenty-five animals were shipped back.

The biggest case of recent times came up at Wall, South Dakota, in the spring of 1949. I heard about it from Earl Gensler, sheriff at Rapid City, with details supplied by Secretary Shorty Rasmussen of the Association and Emmett Horgan, the president. Gensler looks like a Hollywood actor on vacation in his

216

neat, chalk-stripe tan suit, but he is a hard-hitting peace officer who did his full share in putting an extraordinarily bold and clever gang of rustlers behind the bars. Furthermore he is a first-class man at telling his own story, and here it is in his own words:

"We started working on this case, or others like it, something like three years ago. We knew the stuff was being stolen. We thought we knew who was stealing it. But we couldn't get our fingers on a thing.

"The first break we had came when we had a telephone call from around Wall, where the stealing was going on, asking us to check on the license of a car that was being seen a good deal down there. We found that it belonged to a man named Raymond Hickson of Del Monte, Indiana. His two sons had driven it out, and they were in company with a third character named B. C. Bowell. Our report said that Bowell was a bad-check writer, but at the moment he was not wanted.

"All this time I was back and forth trying to find out where the cattle were going—and a lot of them were going some place. I wondered if there might be some connection between the man we suspected and B. C. Bowell, so I went and asked some questions.

"The fellow we were really after was a rancher named White, who lived on a run-down place out in the rough country. I went to see him. No, he didn't know anything about Bowell—didn't know him at all. I went out and asked him two or three times, but he wouldn't talk.

"Then we found out that B. C. Bowell and three others were living in an old trapper shack north of Ward White's. They had a jeep in which they went riding at night—raising the dickens. We went down to see about it, and found them temporarily out. They had taken a load of stolen horses to Preston, Illinois, as we found when we followed a tip down there. Before they left Dakota they had written several bad checks, so we got out a warrant and were about ready to go after them when the money came through. They had sold the horses and made the checks good. This was in September, 1948.

"In order to be ready for the next move, we talked to several

217

people around Wall—alerted them—asked them to tell us if anything happened. The result was a telephone call from Wall that two trucks had arrived from Laporte, Indiana. They were to pick up two loads of stock from Ward White. The tip-off had been a telephone call from B. C. Bowell to the trucking firm. He said his name was Ward Kelly, owner of the cattle, and that they would be handled by his foreman, B. C. Bowell, and Ward White.

"Bad weather held the trucks up at Philip and the truckers had to wait a while. During the delay they heard a lot of talk. People told them not to go down there. Others hinted at a crooked deal. The truckers got cold feet finally, and went back to Indiana.

"They got another trucking outfit to come from Sioux City. But again bad weather got in their way. They had a jeep which they used to scout the range, bought with the proceeds of a previous sale—twenty-three head which they had taken to Rensselaer, Indiana. Ward White and a trucker from Iowa got drunk after that sale and went to jail. Bowell spent his time buying the jeep. If everything had gone right, they would have used it to line up the right cattle to be shipped out, but the trucks from Sioux City arrived before the weather allowed them to get out in the pastures. They rounded up the first cattle they could find so as not to keep the trucks waiting. There were fifty head, all from their near neighbors, belonging to eight different people.

"It was typical of this thing that a number of neighbors had been losing for a long time. George Kennedy says he is out $4,500 worth of stock that he can't account for. One man lost so much that he sold out and left the country.

"Some of the neighbors may have been afraid to move against the Whites, who were very tough citizens.

"The two truckloads of cattle went on into Wall. Bryce Kennedy (son of George) saw them go and called the sheriff's office in Rapid City. The office got in touch with the sheriff, who was on his way home from Plankinton, and near Wall at the time. He set the forces of the law in motion; the Highway Patrol put up a road block on every road; the attorney general was notified, and his men were on the job too.

"The trucks went into the White pasture about 10:00 P. M. and came out about 3:00 A. M. loaded down. The jeep guided them out and came on into Wall. But when the officers stopped the trucks, the jeep was parked on a side street, with the lights off, and the men were no doubt watching the proceedings.

"The situation was a little too much for the examining officers. The trucks had very close bars—hard to see between. It was dark. The only brand the men could check was one belonging to Alfred and Leo Bastian, and it had not been reported as stolen. The truckers came from a reputable firm and were indignant about being stopped. They even had papers in the name of G. Kelly consigning the cattle to the Chicago market, where they would necessarily be checked. It might have been legitimate, and everybody was so bold and forthright about it the officers felt they had to let the trucks go on. And they did. But first they got the drivers to agree to have a little motor trouble when they got to Sioux Falls.

"It was my job to make sure the cattle had been stolen. I went to Verne Wilsey, who had eight or ten head in the bunch, and told him he had some cattle in Sioux Falls. He was very much surprised. 'I guess I better do some more ridin',' he said. 'I've got too much spread on my cattle—from here to Sioux Falls.' Wilsey's brother said when he heard that there was enough evidence to get the men eighty years apiece, 'Eighty years ain't enough. They ought to get life!'

"Bowell was bold enough to try to release the cattle in Sioux Falls. When he saw that the detention was serious, he said he would have to get hold of his boss, Mr. Kelly. Then he skipped out for Chicago, Mrs. White going along with him. They told everyone that was where they were going, so the officers did not look there at first. Finally, when other leads failed, they tried it. The Chicago authorities knew about the fugitives and picked them up at Gary, Indiana. They were extradited and sent back to Rapid City.

"They arrived in November, were bound over, and gave bond. Anyway the Whites did. White's was $5,000 and his wife got off for $1,000. Bowell could not furnish bond.

"They were indicted for grand larceny on eight counts in

one complaint. On April 25 they were brought to trial—pleaded guilty—and drew seven and a half years in the state penitentiary. That is, the men did. The woman was acquitted.

"They were quite a pair. Bowell was the brains of the outfit. He was the one who had told White about the lack of inspection regulations in other states. White in turn had told him a hard-luck story that all his neighbors were stealing from him—so they agreed to steal the stuff right back, and more, too. They'd show 'em!

"Bowell was a smoothie—a salesman and a promoter. He had curly hair, was gentlemanly in appearance and manner, about twenty-nine years old and pretty much of a super-salesman. They said his father was a doctor.

"White was a rougher type. A real cowboy who could do anything with a cow or horse, but useless otherwise. He was a heavy drinker and never did work or do any good to anybody. He had seven children, a wife, and a mother on the home place. Both he and his wife owned land and stock, particularly horses, but his oldest daughter was the manager of the family and attended to a lot of the details.

"They all had some idea of what was going on. One of the younger boys was heard to say, 'They'll never catch my daddy. He's an A-number-one outlaw!' "

Emmett Horgan had a part to play when the case came to trial. One of the best criminal lawyers in the state was hired by the defense. He came to Horgan and wanted him to agree to a light sentence in return for a plea of guilty—all the blame to be thrown off on the Indiana boy. Horgan wouldn't do it.

He told the judge when the case was ready to be decided: "The average herd of the cattlemen in this state is about eighty head. If you let this so-and-so loose, you might as well get ready to take care of everybody."

The lawyer asked for five years. Horgan held out for ten years. The judge split it and gave the men seven and one-half years each.

As a result of the excitement over this case, the Association was able to get some much-needed legislation through the state legislature. Cattle leaving the state obviously had to be made subject to inspection, and now it is legal for any inspector or police officer to stop a truck anywhere, unload it if necessary, and ask for clearance papers.

"When you stop to consider," says Emmett Horgan, "that one big truckload of cattle represents five thousand dollars of somebody's money, you can see why something had to be done."

There are those who think the situation calls for much more direct action. Lee Evans says, "I don't believe in killing anybody, but a little birdshot in the right place might make some of these fellows see the light."

"The best way to stop a local man from stealing," says grizzled old Dee Martin of the Chapman-Barnard Ranch, "is to steal his stuff back. That stops it in a hurry. I was losing some one time and knew where it was going, too. There were several men about who were known to be thieves, but hadn't been caught. I went to one of them and told him I wanted him to do something.

" 'We've just lost fifty steers,' I said, 'and so-and-so has 'em. I want you to steal those steers. I know you're not above doing it, and I promise you you will not be arrested or prosecuted.' He got all fifty and shipped them out, and the other man quit stealing.

"Another time a bunch bought some of the brand I was working with and moved them in right next door, just so they could steal ours. They stole ten head from us, so I stole sixteen of theirs. Oh, how they hollered! That kind always holler the loudest. I stopped this fellow one day and said, 'You remember those ten head we lost last year? Well, I know you got them. So I stole sixteen of yours and what are you going to do about it?' He couldn't do anything about it, and he stopped his meanness."

Henry Bell says that Western cattle raisers need not feel too bad about their losses. There is more theft on the Atlantic Seaboard than there is in the range country because the Easterners have not worked out any real system of protection. The fact remains, however, that the rustler still rides in the West, and adds one more item to the list of ranchmen's griefs.

"T 19. The Rancher and the Government

HE TIDE of the earth's population is rising, the reservoir of the earth's living resources is falling. . . . There is only one solution: Man must recognize the necessity of co-operating with nature. He must temper his demands and use and conserve the natural living resources of this earth in a manner that alone can provide for the continuation of his civilization."

These words from the conclusion of Fairfield Osborn's *Our Plundered Planet* are typical of the moans of many prophets of doom who are warning us that we have almost ruined our country by wasting our natural resources, endangering our watersheds, and depleting our soil.

Nobody questions the need for these warnings.

Nobody doubts that we have behaved like spoiled children in a candy store in occupying and exploiting this continent.

Nobody can fail to be aware, if he is aware of anything, that we have threatened ourselves with economic suicide by foolish denuding of timber lands, by reckless plowing in regions too dry for agriculture, by irresponsibly sapping the soil through overgrazing and one-crop agricultural programs—that the best part of our inheritance is in danger of washing away, blowing away, or being "killed" by its users.

Many of us—perhaps all of us—are to blame, but the man who bears the brunt is the American rancher. He occupies the land which is in most danger. His past history lays him open to the charge of carelessness and selfishness in the use of the public domain. And bad moves on the part of his leaders during the last few years have made the whole industry the target of a campaign of hate which has caused him much grief and anger. Not the least of his troubles today is the fact that many of his fellow Americans think he is a crook.

He steps up to the bar of public opinion with two counts against him:

First, he is accused of selfish exploitation of the range, private and public, and of complete indifference to the resulting erosion, floods, sedimentation of reservoirs, and damage to watersheds.

222

Second, he is accused of wishing to grab for himself huge portions of the country which belong to all the people.

There is just enough truth in these charges to give the calamity howlers and sensational journalists a chance to start a hue and cry. The rest of the citizens, who can not possibly know the background of the trouble, listen to the shouting very much in the mood of a man who is told that his house is on fire. He starts doing things at once, feeling that it is better to get out the bucket and the axe than to waste time asking for proof. Meanwhile the cattleman feels like a neighbor who hurries to join the bucket brigade and is accused of setting the fire.

The root of the trouble goes back to the days when the government was trying to get the land into the hands of the American people—all the people, that is, except cattlemen. No adequate provision was ever made for parceling out chunks of land big enough to set a rancher up in business on the arid plains. The 640-acre Homestead Act of 1916 certainly did not do it. It takes little skill in arithmetic to figure that one out. If a man needs to run two hundred head of stock to support his family, and every cow needs 15 to 25 acres, 640 acres means starvation.

There was the public domain, of course, which the cattleman in certain areas has always had to use. He abused it, too, for it was open to everybody, and if he did not get the grass, somebody else would.

When the whole range country had been seriously damaged by plowing and overgrazing, the government awoke to the peril of the situation and reversed its policy. Instead of trying to give the West away, it began to hold on to everything it could. In 1906, Theodore Roosevelt enlarged the forest reserves and began the active policy of conserving, improving, and enlarging the public lands. Now 405,595,899 acres are federally owned in the eleven "Public Lands" states of the West. The proportion controlled by the government varies from 35 per cent in Montana to 87 per cent in Nevada, with an average of 54 per cent for all eleven.

Supervision of this mighty slice of America is divided among the Departments of the Interior, National Defense, and Agriculture, with nine major bureaus (National Forests, Grazing Districts, Indian Reservations, National Parks, Soil Conservation, Navy, Army, Fish and Wild Life) handling the actual administration through even more numerous subdivisions.

There have always been more or less overlapping, conflict, and variation of policy among these agencies—a fact which has not made the cattleman's status as a tenant any easier. If he was able to get hold of enough land to carry on a large operation, his situation was less precarious. But since the land laws, contrived by Easterners who never understood what the West was all about, gave him no direct route to such independence, he could never spread out enough to gain a real margin of safety. The idea that all cattlemen are big and rich is a delusion. In Colorado, according to Forest Service figures, the average cattle permit for grazing on the Forests is 92 head.[1] The National Cattleman's Association calculates that in 1948 80 per cent of the stockmen ran less than 150 head.

Where the use of public land means the difference between survival and failure, these small cattlemen, at the mercy of the forces of nature and the shifting policies of federal bureaus, suffer more than most Americans from a sense of insecurity. Imagine the feelings of a barber in the city who can do as he likes with three chairs but has a government lease on the remaining three chairs in his shop. Suppose the Department of the Interior tells him what he can do with his fourth chair (including the hours he can use it, the price he can charge, and the number of customers he can accommodate); the Department of Agriculture sets up a different scale of hours and rates for the fifth chair; and the Park Service, by executive order, takes over the sixth chair, reminding him firmly that the rules prohibit barbering inside its boundaries.

The analogy is not exact, because excessive use of barber chairs does not set up an erosion cycle which endangers the rest of the country. But we are talking about the barber's feelings. In such a case he would feel exactly the same as several thousand ranchmen who are operating under similar difficulties at this moment.

[1] *KYW Bulletin* (Forest Service), No. 3 (March 18, 1947), 3.

One other chip off the block of history must be considered before we can understand the ranchman's situation on the public lands. This is the system he worked out before there were any bureaus to control him in his use of pastures which he did not, and could not, own. Essentially it was a rule of "First come, first served," or "Finders, keepers."

In Montana, writes Joseph Kinsey Howard, the great herds "grazed upon millions of acres to which the herds' owners had no right other than the fact that they got there first, a 'squatters' ' right which was given recognition of sorts in 1877 in the curious law of 'customary range.' "[2]

At the southern end of the Great Plains the custom was similar. First comers had first rights, and boundaries were settled by mutual agreement. Furthermore, "the agreements regarding such matters as the partition of the range and the use of water became property rights and were frequently bought and sold as land is sold today. Once a man's rights were determined, they were respected by all. The owners of these rights felt that they had priority claims in any disposition of the lands which they occupied."[3]

"Use" and "right" are still rather close together in the cattleman's thinking. Around Pawhuska, Oklahoma, for instance, practically everybody leases some land from the Indians through the Indian agent. "None of us," says Claude Higgins, "would think of trying to get hold of a piece that somebody else had or that adjoined another man's holdings. Of course, if the man didn't want it, that would be something else. There were feuds in the early days when one man bid five cents an acre more and took another's lease out from under him."

Wherever the cattleman occupies public lands, he attempts to put these inherited notions into operation—and everybody else thinks he is greedy and presumptuous. It is all public property, belonging as much to the citizens of Cicero, Illinois, as to the stock raisers who use it. The man from Cicero cannot see

[2] *Montana High, Wide, and Handsome* (New Haven, Yale University Press, 1943), 104–105.
[3] B. Youngblood and A. B. Cox, *An Economic Study of a Typical Ranching Area on the Edwards Plateau of Texas,* Texas A. and M. Experiment Station, *Bulletin No. 297* (July, 1922), 71.

what right the rancher has to complain when the Forest Service pushes him out. He did not belong there in the first place. The landlord has told the tenant to move, and who does the tenant think he is to talk about his "rights"?

Well, the cattleman has been given a "priority" on his forest land, just as his grandfather was given first chance at his portion of the old free range. Priorities, which mean the difference between a living and getting out of the business, are a form of property, and always have been in the cattle country. Forest permits are transferred with the sale of patented land, and serious grievance results when somehow the permit is reduced or turned over to somebody else. The man from Cicero thinks the rancher occupies his forest permit on sufferance. The rancher thinks he has established a right to be there indefinitely. And if that right is jeopardized, not merely his own economic security but that of everybody else in his situation is imperiled. A great deal of unprofitable argument has resulted, and will probably continue to result, from failure to understand this particular situation.

So the rancher is a more or less unwelcome guest in the national house, who nevertheless has some special ideas about his right to be there. Now let us see how he has abused or respected the hospitality which has been so grudgingly offered him.

From the very beginning he has known that a set of house rules was desirable and necessary, but none was provided. As early as 1900 the National Cattlemen's Association tried to get a federal leasing bill through Congress, but there was so much opposition, even among stockmen, that the legislation got nowhere.[4]

Nothing on a national scale was done to bring order out of this chaos until the passage of the Taylor Act in 1934. This important and revolutionary measure set up an administration charged with the regulation of grazing on the left-over lands which seventy-five years of homesteading had rejected as unfit for human settlement. Districts were organized with the consent and co-operation of the stockmen, who served on local boards and participated actively in the management of allotments.

[4] Dan Fulton, "Range Tenure on the Northern Plains," *The Montana Stockgrower*, October 15, 1949, 15.

There is no doubt that the Grazing Service, which Farrington R. Carpenter put into operation in 1935, was a tremendous
step forward. At last the most abused part of the public lands
had a chance to recover. The Taylor Act "brought about the
first semblance of order on the ranges ever attempted."⁵

In the eyes of the ranchmen it was a great step forward in
another way. For the first time local boards of ranchmen shared
responsibility with government men and had a voice in deter-
mining their own fate. Since that time they have desired and
demanded the same privilege in dealing with other public-land
bureaus, to the infinite horror of the bureaucrats. And although
officials of the Grazing Service and of the Resettlement Administration (created in 1935 and administered by the Soil Conservation Service) are proud of their relations with their advisory boards, the cattleman's enemies still point the accusing
finger. Through these boards, they say, the villains have "cap-
tured" the Grazing Service and are free to continue the ruin of
the range at their own vicious will.⁶

The truth is that even before the Taylor Act was passed,
many ranchmen were working hard to improve the ranges, and
the good ones fiercely resent attempts to lump all cattlemen together as willful despoilers, and to damn the whole industry
with half-truths. They point out that it is impossible to stop
erosion completely, that the range is in much better shape, on
the whole, than their critics admit, that they make their living
from the grass and would be the last men in the world to destroy
it wantonly, that the horrible examples cited against them are
usually hand-picked from parts of the country which went to
the dogs several thousand years ago. It is as easy to show that
they are right as to prove that they are wrong. It all depends on
where you go to take pictures, or which bulletins you read.

"Severe, accelerated erosion—erosion speeded up by human
activity—is occurring on 12,346,777 acres, and nearly one-half
of the State, 30,282,800 acres, is suffering from moderate accelerated erosion,"⁷ say the Colorado soil specialists.

⁵ F. R. Carpenter in *The Grazing Bulletin*, March, 1936.
⁶ Lester Velie, "They Kicked Us Off Our Land," *Collier's*, July 26, 1947, 40.
⁷ T. G. Stewart and A. J. Hamman, *The Soil Conservation District*, Colorado
State College Extension Service, *Bulletin No. 373-A* (July, 1941), 6–7.

Compare this statement: "Recovery of the grassland vegetation following the drought has been comparatively rapid. Blue grama and western wheatgrass, the most severely injured of the major species, are now approximately at the same level they were prior to the drought period. Other major species are of even greater abundance now than they were in the pre-drought period."[8]

Or this: "When the inexperienced man sees the barren ranges like they were in 1934, he immediately concludes that they have been ruined by overgrazing and that Federal control must be increased to protect the stockman against himself. . . . Senate Document 199 in discussing 'The White Man's Toll' showed by a diagram that Wyoming ranges had been depleted 51 to 75 per cent due to the apparent indifference of those controlling the range. . . . Our Wyoming studies show that there has been no measurable change in the carrying capacity of the privately controlled ranges during the past 40 years."[9]

Ranchers in Wyoming and the Dakotas feel that the range was never better. In New Mexico and West Texas, where rain is as scarce as gamblers in church, the grass suffered more complete ruin and will be much harder to bring back. But even in the deserts great progress has been made.

Royal Brinkerhoff, who runs cattle on the Utah desert in winter and on the forest in summer, says: "When I came on the range in 1924, it had been badly used. The permits are not 50 per cent of what they were in 1924, and some spots are still rough, but comparatively the range is in wonderful shape. The eroded places are grassing over—it is just a matter of time, and good management."

Management! There is the word that everybody trips over. Management means that somebody has to do the managing, and in the long run the issue boils itself down to this: Is the ranchman the best manager, or must he be told what to do by somebody

[8] Warren Whitman, Herbert C. Hanson, and Ronald Peterson, *Relation of Drouth and Grazing to North Dakota Range Lands*, North Dakota Agricultural College, *Bulletin No. 320* (February, 1943), 23.
[9] Maurice Haag, editor, *The Range Lands of Wyoming*, University of Wyoming Agricultural Experiment Station, *Bulletin No. 289* (February, 1949), 12–13.

higher up? And that can be boiled down still further to the question of whether the ranchman has the intelligence and the will to protect himself and the resources which he holds in trust, or is incapable of handling his own affairs and must be controlled like a bad child.

The most vocal section of the chorus of critics says positively that he cannot be trusted. Mr. DeVoto, in the passage quoted at the beginning of this book, declares that "Nothing in history suggests that the states are adequate to protect their own resources, or even want to, or to suggest that cattlemen and sheepmen are capable of regulating themselves even for their own benefit, still less the public's."

This is, to put it mildly, an uninformed statement in the light of what thousands of responsible stockmen are doing to take care of their own problems. They have set up, voluntarily, thirty-seven conservation districts covering fourteen million acres on the Great Plains.[10] In Montana, particularly, small ranchers have combined and formed grazing districts of their own, usually in co-operation with the Soil Conservation Service, the Grazing Service, or both,[11] and have thereby set an example which "has changed the nation's livestock practices."[12] Conservation men complain that there is much inertia to overcome in persuading the older cattlemen to adopt intelligent conservation practices, but even the old-timers are farther down the road than the skeptics will ever believe.

Let one example stand for all that could be cited: On his eighty-section Antelope Springs ranch north of Casper, Wyoming, George Snodgrass has built fifty-two dams, many of them equipped with separators or catch basins to trap silt just before it gets to the main reservoir. "You can watch the grass spreading back up the valleys behind those separators," he says.

He has built forty miles of roads on the ranch, mostly on the tops of ridges. When a road crosses a valley, he runs it over the

[10] B. W. Allred, *Range Conservation Practices for the Great Plains*, U.S.D.A. *Misc. Pub. No. 410* (December, 1940), 19.

[11] G. H. Craig and Charles W. Loomer, *Collective Tenure on Grazing Land in Montana*, Montana State College Agricultural Experiment Station, *Bulletin No. 406* (February, 1943), 10 ff.

[12] Howard, *Montana High, Wide, and Handsome*, 300.

top of a dam, for observation has taught him that erosion of the worst type comes from the washing of trails and roads.

George is now sixty-four years old and homesteaded his place in 1907. He fought for it, starved for it, and slaved for it, and more than once nearly lost it. The dugout in which he lived for years still stands on his place as a reminder of those hard old days. While he was building it, he cooked his meals over an outdoor fire while his hands stiffened with cold. George is no product of wealth and tradition. He learned everything by himself, even to the use of surveyor's instruments, and he has helped others as much as he has helped himself. He hates waste, inefficiency, boondoggling, and lying worse than any man I ever saw, and neither he, nor the hundreds of ranchers like him, will ever shovel their land into the nation's rivers.

Everybody needs some correction and restraint, cattlemen included, but there are enough George Snodgrasses among them to keep them at least as responsible as the coal miners, bankers, and potato farmers who have upset the national applecart in ways which no rancher would ever be guilty of. If ranchers cannot be trusted, in the mass, to do their duty, then there is no hope for the country.

All this would not need to be said if it were not for the painful developments which scandalized the nation in 1947—a series of conflicts and crises with the Forest Service over which the blood of both sides is still boiling.

As the drought of the thirties progressed, the forest ranges, already hurt in many places by heavy grazing, began to cause the Service great concern. The men in charge commenced reducing the permits in order to save soil and grass from complete ruin. They were conscientious about it. The cattlemen admit that, and admit also that of all the government bureaus "none are more imbued with a sense of stewardship for a vast national resource than the Forest Service."[13] What they objected to was the way it was done. Some of the cuts, they felt, were drastic and unnecessary; some were made without proper notice; the officials were often arbitrary and insulting; and there was no

[13] Farrington R. Carpenter, "Grazing on the National Forests," *Western Live Stock*, February, 1940, 19.

effective appeal or recourse from their rulings. The Forest Service, they said, was judge and jury.

The supervisors replied that they were thinking and planning not for next year, but for the next century—that pressure on newspapers and legislators from the cattlemen hampered them in doing their duty—that they hated as much as anybody could the hardship and loss of friends that their policies entailed.

"When you've lived with these men and made friends with them," says Ranger D. E. Gibson at Steamboat Springs, "you think twice before you recommend a cut."

The sort of thing the ranchmen complain of happened to Al Favour of Prescott, Arizona, who got a permit in 1936 to graze 1,476 head of cattle the year round on his East Bear Creek allotment. He voluntarily reduced his herd each year for range protection until he was running 583 head. Then he went off to war, after having reached a clear agreement that the arrangement would stand until he got out of the navy. It stood—until he got notice through the mails that he was to be cut 60 per cent. He had to ask for leave, return to Prescott, and fight for all he was worth before he could get back into the war without having his living cut out from under him.

The hottest spots were, and are, in western Colorado, where tempers have been worn to a frazzle for years and the air is a fog of accusations and counter-accusations. The word "trespass" is the one heard oftenest. Cattle left on the range after time is up are said to be in trespass, and their owners are fined. Three offenses are grounds for taking their permits away entirely. The Forest men make all the decisions, and are responsible to no one but their own superiors. They can be summoned into no court but their own. A new five-member National Forest Advisory Board of Appeals, set up in January, 1950, may relieve the situation, but cattlemen with grievances are waiting to be convinced.

Probably the angriest man in Colorado about this is Floyd Beach of Delta. The Forest men think he is a sorehead. He thinks he is a man with a just grievance.

"It just isn't American to do things this way," he says. "It's dictatorship, and we won't take it lying down."

The outcry from the Grand Mesa, the Uncompahgre, and other critical areas brought on the Follies of 1947. It was a play in two acts, the first being a barnstorming attraction headed by Congressman Barrett of Wyoming, chairman of a subcommittee of the Committee on Public Lands. In a series of hearings held all over the West, Chairman Barrett did his best to make a case for the cattleman, and in so doing laid himself open to a charge of bias which damaged, if it did not nullify, the findings of his committee. Some of the sessions degenerated into noisy free-for-alls which some stockmen considered disgraceful. Luke McOllough says he stood the one at Grand Junction as long as he could, and then walked out.

While this was going on, the officials of the cattlemen's associations were making some unbelievably stupid moves. B. F. Davis, secretary of the Colorado organization, was reported by the Denver papers as having said publicly that "the cattleman himself is the one to judge how much and when to cut his herd."[14] *Western Livestock,* organ of the American National, broadcast a call to its members to send in examples of arbitrary and unjustifiable action on the part of the Forest Service to be used in connection with the Congressional investigation—a move which made it look as if the cattlemen were trying to rig the hearings.

Act Two of the Follies was played simultaneously with Act One. Apparently it was devised by a small group of big operators who assembled at Reno, Nevada, in August, 1946, and met again at Denver to work out the details. The story broke when J. Elmer Brock, past president of the national organization and one of its most effective men, aired his views in the Denver *Post* on February 2, 1947. He proposed that the Taylor Lands—130,000,000 acres—be sold to the ranchers who were using them. He wanted to pay for them—10 per cent down and thirty years to liquidate the rest at 1.5 per cent interest. He said it was the only way to bring stability to the industry since the lands were "held in trust and administered by fifty-nine land management agencies, many of which we cannot trust."

At that moment the sky fell in on J. Elmer Brock and his friends. Conservationists, sportsmen, and journalists frothed and

14 *The Record Stockman,* April 4, 1946.

raged. Many local associations publicly declared their abhorrence of the mighty grab. Brock himself is said to have remarked, after nearly being blown away by the blasts which followed his article, that he guessed there were only three men in the country who wanted that government land—Norman Winder (president of the Wool Growers), Dan Hughes, and himself.

The sad part of it was that all cattlemen looked alike to the angry men on the other side. They damned all ranchers, individually and collectively, as would-be thieves and robbers. It was a body blow from which the cattleman has not recovered. The American National has set up a public information committee which is trying a little belatedly to take the bad taste out out of the public mouth, but it is uphill work.

For one thing, the committee has to try to convince the rest of the country that the cattleman is not talking nonsense when he says he is being run out of business by the "cancerous growth"[15] of the government bureaus. It is true, however, that the parks want more land, the wildlife people want more land. The army, navy, and rocket men want more land. And they are getting it. The Forest Service, for instance, added 23,000,000 acres to its area between 1918 and 1948, and the other bureaus have grown correspondingly.

A sore spot is the Tularosa Basin area in southern New Mexico where a total of 1,250,000 acres is in use by two separate groups engaged in the development of guided missiles. In 1948, General John L. Homer stated that $40,700,000 had been spent on the larger of these—the White Sands Proving Ground. By contrast, the penny pinching of the government in dealing with the ranchers who were forced off their holdings in the area seems completely pitiful and unnecessary. True, only a small part of the range was patented land, but the use of government and state leases had enabled these men to develop their own holdings and support their families. When they were forced off, their buildings, fences, and water systems went to ruin, and at the same time they were handed a pittance in no way comparable to the living that had been taken away from them. Columbus

[15] Paul Friggens and Ray Anderson, "Give the Ranchman a Fair Deal," *The Farm Journal*, October, 1948, reprint.

McNatt left a good little ranch and retired to Old Mesilla on $900 a year, and the rest did not fare much better.

This could be borne, since it was in the public interest, but there was more to come. In July, 1948, it became known that the army wanted a thirty-three-mile northward extension of its area, taking in another 689,904 acres. It was estimated by G. W. Evans, president of the New Mexico Cattle Growers, that the forty ranchers who would be displaced would raise the total of affected families to 114; that 40,000 cattle and 30,000 sheep might have to be eliminated—all so that the army could shoot off a dozen rockets a year. It seemed to these stockmen that if some workable system of co-use could not be found, at least they should be allowed to exchange their holdings for land somewhere else or be paid enough so that they could make a fresh start.

The final decision was not reached until November 22, 1949, when the army settled for an arrangement a little closer to justice than anything they had proposed before. A rancher with one hundred cattle, valued at $12,500, will get $1,250 annual compensation from the government, from which he must pay his own rental fees for the leased land. The contract runs for twenty-five years, leaving ultimate title with him. This offer, by no means princely, is three times what was originally proposed, and the ranchers are taking their medicine and signing up. The announcement in December of the creation of a new Air Forces Special Weapons Command with headquarters at Albuquerque and with more millions to spend makes one wonder, however, if the New Mexico ranchers are out of the woods yet.

A bad situation of another type has been near the boiling point for several years in the Jackson Hole country of western Wyoming. It is lovely country, with the wild majesty of the Grand Tetons dominating the landscape, and as early as 1898 proposals were made that at least part of the area should be turned into a national park. After repeated attempts which met with no success, 96,000 acres were finally set aside in 1929. The people of Teton County were assured that if they consented to the withdrawal, no more expansion would be attempted.

Four or five years before this, however, John D. Rockefeller,

234

Jr., inspired by the purest of patriotic motives, had begun to buy up land adjoining the park area, expecting to donate it as an addition to the park. Four times legislation failed which would have extended the park boundaries, leaving Rockefeller with 32,000 acres on his hands. In 1942 he made it plain to Secretary Ickes that unless something was done immediately, he would get rid of his land some other way.

The result was the act from which Jackson people date time. On March 16, 1943, President Roosevelt established the Jackson Hole National Monument by Presidential decree. Now 173,-065 acres in the monument belong to the government. The rest is privately owned but will eventually become public property. Again and again Congress has turned down a bill to abolish the Monument, but the people have formed a committee for the Survival of Teton County and have no intention of giving up the fight.

It is really a serious matter for them. As things stand now, the government owns 93 per cent of the county, says Felix Buckenroth, local banker and powerful supporter of home rule. When all the private land has been taken in, the government will own 97 per cent, none of it on the tax rolls. "We got to make a living," snorts Mr. Buckenroth. "*Somebody's* got to make a living. The Grand Tetons looks just as good to me over the back of a cow!" Already, says rancher Cliff Hansen, the taxable land is assessed at four times the state average for tax purposes.

The cattlemen, who drive 16,000 head of stock back and forth across the Monument twice a year and would like to use the land for grazing, are the ones who stir up most of the fuss. It is hard to tell them that they do not have a grievance.

So it goes all over the West. The great dam projects on the Missouri and other rivers may benefit posterity, but many cattlemen will be put out of business, and they have to decide whether they will be patriotic and out of luck or scream like the devil and be called obstructionists who stand in the way of progress. Some of them scream, and some don't. "Few cattlemen can think in terms of a hundred years," says Marcus Snyder of Billings, "but I can. Those dams should be built. The population is growing, and all that land will be needed in a generation or two."

In one respect Marcus Snyder is certainly right. The average ranchman, particularly if he is over forty and has not been to college, is a practical man who knows very well what he is looking at, but can't see over the hill. His economics are short range and he believes in taking care of himself first. He thinks that the world would be a better place if everybody would do that, and maybe it would.

To his way of thinking, the whole matter is simple—run the country as if it were a successful ranch and things won't get far out of line. He sees no reason why economy in national affairs should be less desirable than in his private business. He thinks that debts ought to be paid, whether one man or one million men owe them, and that there is no sense in running up bills unless one knows how he is going to pay. He respects production and thinks it should be profitable, despises waste, log rolling, feather bedding, and leeching, and can't understand why anybody practices or puts up with them. He thinks a man ought to be able to do his work without being supervised and does not understand the need for co-ordinators who check on the inspectors who check on the operatives. Nobody talks about bureaucracy with as much bitterness as a cattleman. Nobody values independence and despises collectivism as thoroughly as he does.

To show how completely he rejects such ideas, the American National, with the full backing of most stockmen, has declared itself against any price supports for livestock. It is true that the organization does not like competition from Argentine beef and favors some sort of tariff security, but it sees no paradox in asking for protection from foreigners while rejecting interference from the folks at home.

The policies of state and national cattlemen's organizations are mostly extensions of the ideas of the individual rancher. These groups are out to "protect the interests" of their constituents—that is, get as much for them as they can. They are dominated by rich and powerful men and they are frankly and proudly selfish, as all organizations must be which work for the welfare of a special group. The trouble is that they are too open about their aims. They do not hide behind a screen of verbiage and pretend that they are being persecuted and exploited. They have no John L. Lewis to turn purple over the wrongs of the poor cattleman, and they do not want one. As a result they are considered by many honest Americans as dangerous and unscrupulous in the last degree.

If honest conservatism is dangerous, the cattleman is certainly a menace. He will seldom have any truck whatever with New Deal or Fair Deal ideas, and is even prepared to make serious financial sacrifices, sometimes, for the sake of his political conscience. Lee Evans says he loses $30,000 a year on his 125,-000-acre ranch west of Albuquerque because he won't let the government help him "improve" his range. Bernard DeVoto is talking about such men as Lee Evans when he speaks of "the primordial and obsolete savagery of Western thinking."[16]

Simpler citizens suppose that DeVoto is right. At the 1949 meeting of the Wyoming Stock Growers a pert Denver brunette was keeping the stenographic record. I asked her what she thought of the assembly and she replied pithily with a toss of her bangs, "Twenty years behind the times!"

The thoughtful ones realize the hopelessness of their position. Oda Mason, rock-ribbed conservative and pillar of the associations, puts the realization into words:

"This may be progress. It might be. I suppose our fathers and grandfathers felt this way about the changes in their day. Our children will get along better than we do with this regimentation, and their children will like it because they don't know anything else. It may be progress, but dammit, I don't like it. When people have been living as long as we have under a system of free enterprise, they can't—they just *can't*—put up with this sort of dictation!"

It is not in their blood, either, to throw in the towel when the fight seems to be lost. This battle is not over yet.

[16] Review of *Rocky Mountain Cities*, New York *Herald Tribune Weekly Book Review*, April 10, 1945.

PART V ...but It's Not All Grief

O 20. . . to Become a Useful Citizen

NE CHILLY MARCH EVENING in 1949 four hundred people assembled in the ballroom of the Student Union Building on the campus of the University of Wyoming. They were 4-H Club leaders, county agents, home-demonstration workers, and other delegates—all with a healthy suggestion of the country about them, and all in a mood to rub one idea against another during the annual RFH Week.

The attraction at this session was a sturdy blonde Scandinavian girl from Lusk. Anna Hansen was the daughter of a Norwegian immigrant who had taken up farming and cattle raising in eastern Wyoming and had scraped together enough money to send his daughter to the university. And Anna had done better than even her parents expected. The four hundred people trickling into the auditorium were there to hear her speak.

Serious-faced and a little anxious she watched the audience come in. There was A. E. Bowman, director of the Extension Service, spare and elegant with white hair and tightly clipped mustache. There was Burton W. Marston, state 4-H Club leader, hurrying as usual. There were presidents and sponsors of every agricultural organization in the state. They had backed her; she owed her success to them.

After her introduction, she got up and told them simply and clearly about the places she had seen and the new friends she had made—for Anna was the delegate from Wyoming to the first International Rural Farm Youth Exchange. She was one of the seventeen young men and women from all over the country selected to go abroad as unofficial ambassadors from the common people of America to the common people of Europe.

241

Imagine what it meant to those inland youngsters: Washington—the docks—the steamship—the bustle of foreign ports—the new friends in Belgium and Denmark, Finland and Holland, Norway and England. And it had not cost the government a penny. Contributions from individuals and groups had made up the thousand dollars each one needed for expenses.

Anna told about the Norwegian Agricultural Clubs, started by men and women who had got the idea in the States; about the national health programs; about the spinning and weaving and knitting; about the beautiful furniture; about the ration cards; about the political parties; about the people in Norway and England who seemed like old friends now.

"Washington attachés can tell you about Norway," she said, "but I can tell you that you must live, work, and play with the people to really know."

Anna and her sixteen fellow delegates were the first wave of country boys and girls sent to chip away at the barriers of space and custom and language which keep the good people of the earth from knowing and appreciating each other. A second and larger wave was sent in 1949, including another Wyoming girl—Ruth Harris of Cheyenne. In Montana, Ruth Fenske, who went to Sweden in 1948, is making talks and showing slides. Her friend Rhua Slavens sailed for Belgium in June, 1949. A few young people from abroad have come to this country to complete the "exchange," and more are expected.

The Farm Youth Exchange is just one sample of what is being done to transform the lives of our farm and ranch children. The neglected, underprivileged, undereducated country boy of two generations ago has really come out of the brush. Half a dozen agencies are working to keep him in the country, make him proud of his job, and step up his efficiency. The wonder now is that he ever thinks of any other kind of life even for an instant.

Almost from the moment of his birth he is constantly reminded that agriculture is the basic occupation, that it offers health, security, and independence beyond most vocations, and that his parents, his teachers, and his government are eager to give him the know-how which will make him a success.

A revolution in the lives and opportunities of the boys and girls on the American land has actually occurred. The movement has grown like a snowball rolling downhill. In Wyoming, a leading cattle-raising state, membership in the 4-H Clubs (the "H's" stand for Head, Heart, Hand, and Health) has grown from a little more than 1,000 in 1928 to 4,500 in 1949. Texas has 110,000 members. Membership for the whole United States is said to stand at about 2,000,000. There are Indian 4-H clubs on the reservations and Negro units in the South. City people can hardly realize what a tremendous thing this club work is.

What the lodge is to a businessman, what the U.D.C. is to a Southern lady, what the Boy Scouts are to a town kid, the 4-H and FFA (Future Farmers of America) are to the country boy and girl. But they add one element which no similar organization can boast of. In what other fraternity do the brothers and sisters help each other to earn a better living?

The Smith-Lever Act of 1914 set the wheels to turning. It provided federal funds for an extension service "to aid in diffusing among the people of the United States useful and practical information on subjects relating to agriculture and home economics, and to encourage the application of the same. . . ." Young people's organizations were already in existence here and there in the rural areas—in Iowa, in North Carolina, in Texas—and many of these merged with the new 4-H clubs which resulted from the Smith-Lever bill. They were set up to include young people between ten and twenty-one years of age who organized themselves into groups, selected their own officers, and conducted their own monthly or semimonthly meetings. The county agent was to act as their official supervisor, but much of the responsibility was to be turned over to adult local leaders, usually substantial citizens, who served without pay for the sake of the young people.

Of the boys and girls nothing was expected beyond a little enthusiasm, cheerful co-operation, and the completion of a piece of work which would teach them better methods of agriculture or homemaking.

The director of extension in each state assumed responsibility for the whole program. He was to be a member of the staff

243

of the state agricultural college; his workers (the county agents and county home-demonstration agents) were listed as members of the A. and M. staff. The money to pay for supplies, salaries, and other basic requisites was to come from the United States Department of Agriculture with some bolstering from state, county, and local sources as the need arose.

The program continues as it was originally set up, and has proved to be elastic enough to fit itself to local needs and special circumstances. It has been one of the most quietly successful social experiments ever tried in America. When the leadership is weak, of course, the program is feeble. But when conditions have been right and the leadership strong, the results have been impressive.

The FFA stems from the Smith-Hughes Act of 1917—the National Vocational Education Act which provides federal aid for courses in vocational agriculture. The United States Office of Education does the supervising. The Future Farmers were set up on a national basis in 1928, and have boomed since the close of the last war. Chapters may be organized in any high school where vocational agriculture is taught, the Vo-Ag teacher serving as adviser. A wide variety of recreational vocational, and cultural opportunities are provided. FFA bands and choruses are organized for state conventions. The Rugby, North Dakota, chapter has its own library and librarian. Every boy may learn leadership and co-operation if he is so minded, and there is enough ritual and degree work to keep the restless ones interested. Members even wear special jackets with a distinctive emblem—no insignificant bauble of a fraternity pin, but a glowing and expansive creation about the size of a small pie.

The objectives of the 4-H and FFA are approximately the same. Projects are similar, including anything that a youngster can learn to do better around a farm or a ranch. Feeding programs are favorites with ranch boys and girls in the 4-H groups (the FFA's are exclusively male), though town youngsters are often as much interested. Calf projects, which involve feeding a young animal until it becomes baby beef, are most common, but other fattening and breeding programs may be followed.

The project is completed when the animal is shown, judged, and sold at some specified stock show.

The girls feed, too, though the boys do the bulk of the business. Usually one or two fugitives from cooking, sewing, and homemaking projects will turn up with a calf in every club. They like to handle animals, and enjoy showing the boys that they are superior players in a man's game. C. G. Staver of Colorado A. and M. says they do it because they "can't lead a chocolate cake into the ring." Just before the big moment, the girl feeder gives the long bob or the halo another going over, gets into some attractive ranch clothes, adjusts her smile, and makes those judges think twice before turning her down.

Female competition does not worry the boys half as much as the difficulty of getting good calves. They cost money; and there is no use competing with inferior stuff. If a boy's father has fine stock—well and good. If a public-spirited purebred man will quote a rock-bottom price, well and good again. The boy with nothing of this sort to fall back on has to start looking; and as a result you can see boys, fathers, and county agents scouring the country for suitable material at the right time of year.

So earnest is their search that they sometimes make nuisances of themselves. Josef Winkler learned this the hard way when his Castle Rock Shorthorns won the Grand Championship at Denver in 1947. He had one hundred letters asking for calves, and dozens of telephone calls. One man called him from Alabama at four in the morning.

"I had to turn them all down," Josef says. "We have a small herd and we show in carload lots, so we have to keep our stuff together."

Company ranches like the Matador, with boards of directors watching for leaks and extravagances, quite consistently join Josef in staying out of the 4-H business. But hundreds of herds, big and little, send out a constant stream of young stock to junior feeders. The Yearwood Ranch near Austin, Texas, is one which makes special prices on club calves. Alfred Collins of the Baca Grant ships out eight hundred to one thousand club calves every year. Such men are willing to contribute because they like to help improve the breed, and because they love to be generous.

245

The local show, where a boy gets recognition from his own friends and neighbors, is what keeps the competition going. But the winners who go on to state, regional, or national competition provide most of the glamour.

Probably the best-publicized club in the country is the 4-H unit at Fort Stockton, Texas. Amazing things have happened at Fort Stockton. With the help of the local American Legion post the club has obtained a surplus army van to haul its prize stock, and members make trips to shows as far away as Monticello, Iowa, and Chicago. The boys carry bedrolls, radios, and even a Negro cook. A number of national magazines have told their story.

In 1948 a round-faced, energetic member named Sim Reeves, Jr., had a calf which won Grand Champion at Dallas. Burris Feed Mills bought it for $3,000—and gave it back to the club. The next stop was Chicago, where Sim's calf won Reserve Champion and brought $5,160 at the sale from Glenn McCarthy, the Houston oilman and innkeeper. He bought it to serve to his guests at the opening of the Shamrock Hotel on St. Patrick's Day, 1949. To make the gesture more sweeping, he had several pairs of boots made from the hide for the Fort Stockton boys and invited them to attend the grand opening at his expense.

Then in December, 1949, Roy Bean, the club steer, was declared Grand Champion at the Chicago International and was bought for $13,800 by the Dearborn Motor Company. At $11.50 a pound Roy Bean pushed the all-time price record up another notch.[1] Club members like Genell Slaten get all quivery when they tell about these doings.

County Agent Gallman at Uvalde, Texas, has another good story to tell about a redheaded boy named Herman Kuchner.

"I arranged for him and two other boys to go over to Jess McNeel's place at Fowlerton forty miles the other side of Cotulla to buy a Brahma calf apiece. Jess said he wouldn't sell to those kids. He'd *give* them each one. Since Herman was redheaded, he gave him a red one. It was wild as a polecat—never been

[1] It was pushed up still another on February 3, 1950, at the Houston Fat Stock Show when McCarthy bought the Grand Champion steer for $15,500—$17.30 a pound.

handled. But Herman soon had it to where it would follow him like a dog. He could even ride it.

"He was so fond of it he didn't want to sell it—said he *wouldn't* sell it. Well, sale time came closer and closer and he got more uneasy about it. But they persuaded him to go through with the deal, and when it came time for him to go into the ring, he *rode* his calf in—great big tears rolling down his cheeks. Jess McNeel bid the calf in for $500 and gave it back to him. That kid still cries every time he thinks of it."

There are dozens and maybe hundreds of men like Jess McNeel and Glenn McCarthy who believe that nobody can do too much to stimulate and advertise these rural youth organizations. The late Jim Duncan of Wichita Falls, Texas, was one. Thanks to the Duncan Trust, which he founded, the 4-H Club has $1,500 in the bank, a $1,000 bond, and a $450,000 club barn. Max Carpenter, the county agent, and his assistant, Bill Pallmeyer, speak of Mr. Duncan with a sort of hushed reverence.

It all adds up to the fact that 4-H Club work is turning into big business. For some of the more fortunate members it can turn into a good livelihood. Winners ordinarily sell in the ring after the show for prices above the market, and an ambitious youngster usually puts the money back into livestock. Before you know it, he has the beginning of a herd, and by the time he is ready to go to college or get married, he may be well ahead of the game. At the same time he has learned the business and is fairly well prepared to step into the little world where the conformation of the root of a bull's tail can mean the difference between $25,000 and not much of anything.

Claude Means of Gunnison, Colorado, now twenty-one years old and a Clubber since he reached the minimum age, is one who made the grade. Starting with $150 of borrowed money, he has sold breeding stock worth $7,050 during the last eleven years, and owns at this moment twelve cows, three bulls, and four heifers valued at $5,000. They say he could have made more, "but he didn't want to go too fast."

But the whole story is not told in dollars. Records at the state 4-H Club office show that Claude completed fourteen projects, made sixty-four talks, and wrote forty-seven news stories during his club career. He has served as president, vice-president, and junior leader of his club and has won the un-qualified admiration of his elders. Floyd Betts, the county agent at Gunnison, says of him: "He is gracious in defeat; honorable in his transactions with all his fellow 4-H Club members; is hard working and industrious; is a splendid teacher to younger mem-bers; and believes wholeheartedly in 4-H principles and work . . . the most outstanding 4-H Club member I have ever known."

A more exciting way of making a profit out of club work has been taking hold all over the cattle country and points adjacent —"catch-it-and-you-can-have-it" contests, known in the north-ern part of the range as "catch-it-calf" competitions, and in Texas as "calf scrambles."

These semi-athletic contests, which provide good spectator entertainment at rodeos and stock shows, as well as furious rivalry among the contestants, originated at Chicago in the thirties and spread from there. Local rules vary, but the general idea is to turn about twenty young huskies loose in a ring with half as many calves. Each boy has a rope halter, and if he can get it on a calf and drag his prize across a finish line (in some places he "leads" it out of the ring), he is in the cattle business.

The rules say that no boy can touch a calf while another is working on it; but if the frightened animal gets away, the nearest or the fastest youngster can try his luck. The calves are often strong and active enough to give two or three boys a workout. Nobody has ever been badly hurt at one of these scrambles, but everybody usually goes home with a few hoof marks on him.

And then the real work begins. The regulations of the Denver Stock Show include a fifteen-item set of rules, including require-ments that the calf must be insured, that the boy must write a monthly letter to the donor (a copy going to John T. Caine, man-ager of the show), and that he must keep complete records of costs during his feeding project. At the end of the year when show time rolls around again, he brings his calf back for com-petition and sale.

Does anybody object to so much profitable activity among the clubbers? Yes, many a top organizer does. "We don't want any Chamber of Commerce publicity," says Dr. I. P. Trotter and his associates at Texas A. and M. "We want to emphasize the fact that the program tries to use the things that make up the life of a boy or girl—that the development of the individual is the aim, not the winning of prizes."

Ruth Shepard, associate state 4-H Club leader in North Dakota, feels the same way. "We try to emphasize the local shows," she says. "We believe that the closer home you do these things, the more you sell the people on the program."

The whole question was given a national airing recently by Ladd Haystead under the menacing title "Jackpotitis."[2] Too many youngsters go into competition, he believes, not for the glory, but for the money in it. "In the Northwest to combat this trend they no longer rate individual animals but put all the top animals in a class and give a purple award to all owners instead of the usual blue, red, and so on. Kansas has banned cash awards, and the 4-H headquarters is actively at work to return the competitions to an educational basis."

Mr. Haystead says he expected to be overwhelmed by a storm of protests over this blast, but instead he was flooded with mail from people who agreed with him.

There are many signs that this article touched a sensitive nerve. One is the beginning of a movement away from calf scrambling. At Rapid City, South Dakota, where the 4-H program is going like a rocket, catch-it calves are out. Over three hundred calves are shown by club boys and local ranchers at the annual Calf Show, and the competition is keen, but instead of money the winners are offered a mouth-watering variety of trips and excursions. Awards are made on the basis of a point system which gives credit for club activity as well as for success in the show ring. Under this system girls have as much chance as boys and are not pushed to the sidelines as they were at calf scrambles. Likewise, in Montana, the idea of trips and scholarships is gaining ground, and every year ten outstanding members are sent to the National Club Camp at Washington, D. C., to the American Youth Foundation Leadership Training camp at Shelby, Michigan, or to the annual Junior Club Week convention at Olds, Alberta. This last is a pleasant hands-across-the-

[2] *Country Gentleman,* June, 1949.

border gesture which the Canadian students repay once or twice during the year.

In Texas, calf scrambling holds its own, but it has had to surmount a very unusual obstacle—the charge of athletic professionalism. Some smart coach found out that a number of high-school athletes were getting away with prizes at the Houston Fat Stock Show and other places and argued that this fact disqualified star players on a rival team. Coach Claude Everett of Meridian says the calf scramble at the Bear Club Rodeo at Waco in the fall of 1947 touched off the fireworks. High-school football in Texas is serious business, and some of the winners at the Bear Club event were on teams which had a chance at the championship. The wheels started moving. Joe Daniels, who later scored the last touchdown in the game between Gatesville and Meridian, had won a calf, but got scared and gave it back. Several other players were declared ineligible.

Consternation spread like a ripple from Roy Bedichek's Interscholastic League office in Austin. There was much worry and argument. Eventually a scheme was worked out which made calf-scramble winners custodians rather than owners of their calves. They were charged with the care and feeding of a valuable animal, and should, of course be paid for their work. As payment for services rendered, they were given official title to the animal when they brought it back for exhibition at the next show.

Admitting all the dangers and penalties of "jackpotitis," it would seem that not one club in a hundred is in much peril. Accomplishments?—Yes! Gold-digging?—No!

As a single sample of what united boy power can do, take the case of the 4-H clubs around Torrington, Wyoming. Under the guidance of hard-driving County Agent B. H. Trierweiler, they have completed a concrete-block clubhouse forty by one hundred feet with a stage and dressing room at one end and a kitchen at the other. It cost more than $13,000, but the club owes less than $700 on it now. All sorts of conventions and meetings are held there, and the enterprise has been so useful and so successful that the builders have no regrets when they think of the dances, pie socials, cakewalks, sales, and rodeos they sweated over in order to raise enough money to get started. This sort of money raising is not a symptom of "jackpotitis."

Constant effort to improve the project work and fit it to specific needs is an even better corrective. Out in the arid range country of western South Dakota there is little chance for boys and girls to grow plants or even raise hogs or poultry, but there is much that they need to know about their own environment. With that in mind, Henry P. Holzman, associate animal husbandman with the Extension Service at Rapid City, has set up a project in range management of which he is justifiably proud. Aided by District Club Agent S. D. Ham, he encourages the boys to recognize, cultivate and evaluate the grasses on their home range. To complete the project, they have to gather and mount specimens, arranging them in order of their economic importance. The same thing is done in reverse with noxious weeds.

Best antidote of all for jackpot trouble is a conscientious attempt to put into operation the broad objectives of the 4-H and FFA. Properly carried out, they present an almost complete blueprint for developing country boys and girls into useful and effective citizens. The young people learn how to conduct meetings (with pointers from the United States Department of Agriculture on parliamentary procedure), enjoy singing from their song sheets, report club news for the local papers. They keep careful records, not merely of their projects, but of meetings, demonstrations, exhibitions, recreations, accomplishments, and even general health. In the South Dakota Member's Record is a sheet on health habits with blanks to be checked every day beginning with "Have happy cheerful disposition" and ending with "Have bowel movement daily."

There is, besides, a good deal of ritual and ceremony: forms for opening and closing meetings, an Achievement Day at the end of the year to honor faithfulness and accomplishment, the annual Rural Life Sunday which gets everybody and his parents into the churches for a round of preaching and hymn singing as a means of adding a certain solemnity to the year's activities. At Montana State College there is even an honorary Greek Letter fraternity (Mu Beta Beta) for 4-H Club members on the campus.

251

At encampments and assemblies many subjects are offered from handicrafts to discussions of world problems. In such states as Montana and North Dakota there are even conservation camps every year for boys who have been active in soil conservation and want to learn some more about it.

And the contests! You wouldn't believe there were so many opportunities for competition if you didn't see a printed list. The *Wyoming 4-H Club News* for April, 1949, gives details of forty-five state contests and twenty-five national contests open to members with prizes whose combined value runs into many thousands of dollars. The *4-H North Dakotan* lists possibilities all the way from a Foley flour sifter to a $500 award for a top corn raiser, with $200 college scholarships scattered about liberally in between. Donors include Sears Roebuck, Flax Utilities of the United States, the Milwaukee Railroad, and many others.

The boys and girls who are eligible to compete for these prizes come from the widest possible variety of backgrounds. Not all of them, by a long shot, are country kids, though they are necessarily boys and girls who love the country. Seventy boys are enrolled in calf clubs in Shackleford County, Texas, under supervision of W. C. Vines, and not one is the son of a ranchman. Their fathers are farmers and pumpers from the oil fields. Vines explains that 50 per cent of the county is owned by ten men, and those ten men are either childless or can't seem to produce anything but girls. He is frankly worried about this failure of the old stock.

Of course the inclusion of every type of youngster means that there will be a few bad eggs in the basket. Bill McPhee, of Horse Creek, Wyoming, was still snorting about one of the bad ones when I met him outside the Sportsman's Club on the Fair Grounds at Sheridan during the Stock Growers' annual meeting. Bill drove up in his battered pickup, took his spare tire out of the back and locked it up in the cab, and headed his boots, levis, and unreconstructed personality for the chuck line in the clubhouse. I was just ahead of him. We got to talking about cattle stealing, and went on from there to the larcenous habits of the public in general. Bill was unhappy about the moral standards of some of the youngsters he knew. He is a 4-H Club leader, and had just returned from taking his boys to Cheyenne. Back on the ranch, the morning after their return, one of them pulled out a complete place setting of Plains Hotel silver.

"I busted him right in the mouth," Bill declared, "and I made him take the stuff back and apologize. I just had to do something. By God, if we're trying to teach these boys . . . ." He broke down and reached for his Sunday language. " . . . if we're going to make good citizens of them, well, we can't have that sort of thing happen. And when those fellows invited us in and were so nice to us, too. The hotel man said he lost twenty-five sets like this one. There were two hundred people in the party, And he got only the one set back."

So it isn't all sweetness and light. But there are many times more good things than bad to talk about.

The quality of livestock goes up where 4-H clubs and FFA's are strong, says Louis Franke of Texas A. and M. College. Intelligent soil building and good conservation practices are on the increase where the youngsters learn better methods from trained leaders. If our plundered planet can be saved by education in its race with destruction—if the people, given enough light, will actually find their own way—then youth programs in this country are doing more than the bureaucrats in Washington to make sure that our people will be here, and making a living, a thousand years from now.

The program gets back to the parents, too. Herbert Schroeder tells about one substantial but stand-offish ranchman at Columbus, Texas, who sneered at the whole idea until his son started feeding a calf. Then Father got interested. He was especially impressed by the calf's head, chest, and forelegs, and did considerable bragging about the fine animal they had. The calf went to the Houston Fat Stock Show and showed up well until the judges turned the stock around and lined them up from the rear. Father's pride went down with a crash.

"Goddam," he remarked. "We'll work on the back end next time."

A few ranchmen and farmers are still doubtful about the advice given by armchair agriculturists in the county agent's office. If they would listen to J. P. Perkey, director of vocational education at Oklahoma A. and M., the last of their doubts would vanish. Perkey is a large, rough-hewn, harassed man, looking more like a football coach in the middle of a hard schedule than like an Apostle, but an Apostle he is, preaching the gospel of youth organizations with a rough-textured eloquence that is hard to resist. Over an empty coffee cup and a mountain of papers he launches forth:

"The FFA isn't just a club. It's a great idea. The object is to establish boys in farming and ranching—get them ready to support themselves on the land—make them contented with their occupation. We try to build a lot of pride into these boys. And remember that a contented life these days includes a good many things besides a ranch house and a herd of cows. Good roads, electricity, and everything else that makes life good comes into the program.

"We don't want to promote one particular skill—pecan raising, or rabbit culture, or stock showing. We want to build all-round competence as agriculturists."

He points to a recent bulletin on "Activities and Accomplishments" which shows that Oklahoma FFA's now own $4,000,000 worth of land, livestock, and equipment in the state, that in eighteen months before the middle of 1949 these boys constructed 1,000 miles of terraces in their soil-conservation program. And so on.

How much do the ideas and principles of these youth organizations have to do with the conduct of the members in later life? There may be some legitimate doubt about what the results are. In a country which looks first at a man's bank account and then at his character, is all this talk about responsibility, leadership, citizenship, and so forth, just so much hot air?

Who can say? The United States Department of Agriculture has no statistics on such matters, but these things we know: that what a child hears from a person he respects, he believes; that repetition is the mother of memory; and that no vivid impression is ever completely lost. In short, this training for citizenship and character cannot possibly be wasted, and may be of tremendous importance.

Governor Roy Turner of Oklahoma is a good man to say the

WIND-UP OF A TURNER FIELD DAY
Governor Roy Turner of Oklahoma (left) and the
Guthrie, Oklahoma, FFA livestock judging team,
winner in 1948 over 174 competing teams. The
others are Donald Coffin, Dave Morrisett, Donald
Thompson, and Vo-Ag Instructor, Ralph Dreessen

final word. He was a farm boy himself but never a 4-H Club member. "I came along too early," he explains. But he worked so hard with these organizations that many Oklahomans think his 4-H contacts put him into the governor's mansion. Every year in June he stages a mammoth Field Day at his Sulphur ranch attended by as many as two thousand boys and girls. Several classes of stock are shown and the young people compete in a judging contest, the winner walking off with a handsome plaque. At noon they dive into a lunch which costs the governor several hundred dollars, but he considers the money well spent. He believes that if these youngsters are allowed to see good stock, they may develop a desire to eliminate the scrubs from their own herds. To make sure that he has done everything possible, he sells calves to club members at approximately half price—$75 for an animal that would bring $100 to $200 on the market. "I can come out on it," he says.

I called on him in his office to see if I could find out how he really felt about club work. I met a handsome man with smooth gray hair and a friendly manner who sat completely relaxed in his brown leather chair while I asked the final question.

"Has the club program lived up to your expectations."

"Yes," he said. "It has become of far greater importance than I ever expected or hoped."

"I 21. Boots in the Classroom

'D LIKE TO KNOW why in hell *my* son can't go to school!" The startled Dean looked up at a ranchman with a sandblasted face under a battered Stetson. Alongside him was a junior model of himself who looked interested but kept his mouth shut.

"Well," said the Dean, "maybe he can. What seems to be the trouble?"

It turned out that he could. Texas A. and M. had turned him down on some technicality which was not part of the machinery of Texas College of Arts and Industries at Kingsville. He was admitted, and is still going strong on the widespread, breezy cam-

255

pus, made over from a wartime navy base, where South Texas boys learn how to be agriculturists instead of cattlemen and dirt farmers.

J. W. Howe, who runs the division, overheard the conversation and said to himself that things had certainly changed since his days at the University of Alberta. Book learning, once sneered at, is now in demand. Once your old ranchman demanded to know why the hell his son should. Now he asks why the hell he can't.

The old-model ranchman still turns up occasionally, and sometimes in unexpected places. Bob Donelson, for instance, who operates on 16,000 acres at the base of the Flint Hills near Pawhuska, Oklahoma, is definitely a man of the world—star polo player, chronic traveler, inexhaustible sportsman, member of the Osage tribe, and prosperous cattle raiser. He sent his son and daughter to high school in the little town of Burbank, and when they had completed their secondary education, they became ranchers like their parents, in the sturdy tradition of acquiring practical knowledge, the only way some men hold it can be learned—on the land itself.

"We wanted them to have the best advantages we could give them," says Bob, "and I shouldn't wonder if country living doesn't have its points after all. You can't confuse means and ends out here on the prairie."

Not many cattlemen feel the way Bob does. They can afford to send their children off to college now, and a great many of them are doing so. Numbers go to liberal arts schools—there are no statistics to show how many—but if any institutions of higher learning are going to affect the ranchmen of tomorrow it will be the A. and M. colleges with their far-flung ramifications of research projects, laboratories, field days, experiment stations, extension courses, traveling specialists, and mountainous stacks of bulletins. Of all the collegiate centers in this country, they are probably most thoroughly keyed to real needs, most conscious of a mission, most productive of tangible results.

For one Andy Anderson who brings the flavor of Princeton to Encampment, Wyoming—for one Francis Warren who leaves Cheyenne for a couple of years at Harvard because his father and grandfather went there—there are a thousand Joe Smiths who put on their best boots and hike themselves off to College Station or Bozeman or Fort Collins to learn ranching from the ground floor of the Agriculture Building up. When they graduate, they go away to a great variety of jobs, proud of their sheepskins, as well trained as can be expected, and bearing as impressive a load of general culture as most graduates of liberal arts colleges.

It was not ever thus. In fact, it was not even close to being thus a few years past. Mr. Howe feels that a good deal of stiffening has been necessary in building up standards. Not so long ago a course on the Ag campus was a "crip" for football players in need of some easy marks, and even bona fide Ag students expected a maximum of grades for a minimum of work.   ·

At Oklahoma A. and M., Dean Blizzard says the last fifteen years have seen a tremendous upswing in seriousness and interest on his campus. At North Dakota the story has been the same. "We flunk them without hesitation now," says Dean Walster. "They're pretty well staying away from us unless they mean business."

As with all colleges, the big push has come since Hiroshima. The eagerness of veterans, with four years of time to make up, has been partly responsible. But changes in attitude on the part of college administrative officers has probably been just as important. The North Dakota school started moving, says Dean Walster, when the Animal Husbandry Department became interested in the problems of the people of the state. "Our range country is four hundred miles west of here across the Missouri River. Between 1913 and 1941 one of our men got as far west as Valley City. The smartest thing this institution ever did was to set up the Experiment Station at Dickinson. When we began to serve the people, the people began to support us. We go to almost all the stockmen's meetings, and the stockmen come in here to see us. The men who call on the legislature en masse to support the college program nowadays are the livestock groups."

The demand for specialized agricultural education is so strong that courses and departments are multiplying like guinea pigs. Junior colleges and teachers' colleges are getting on the band wagon in a wild rush. The older and more solid institutions watch the scramble with deep misgivings. Agricultural courses are expensive to teach and cannot be properly set up on a shoestring. The standing of all the Ag schools will be damaged if too much low-quality work is offered.

But at least the country is interested—and it ought to be. The Ag colleges have much to contribute to the progress of American education. The campuses at Montana State and at Utah State are among the most beautiful in the nation—tree shaded, mountain ringed, sumptuously landscaped. Course offerings are so broad that an interested student can study anything he wants to, from Milton's poetry to oboe playing in the college symphony orchestra. Liberal arts courses are always available, and degrees are offered in subjects not connected with agriculture at all. In short, a good agricultural college nowadays is a college first, and can foster almost any special interest, with the possible exception of Sanskrit, volcanology, or the Provençal poets.

Although the curriculum is remarkably uniform—almost standardized all over the country—changes are being, and need to be, made. The Association of Land Grand Colleges sees to that. In fact a curriculum committee is working on suggested changes for the Association now. And meanwhile most schools are trying out new ideas on their own.

The course for the first two years is the same for everybody. English, basic sciences, some sort of mathematics, introductory courses in agronomy and animal husbandry, economics, and military science—those would be the fundamental requirements. The experimenting comes in with various additional subjects. Texas A. and I. has added farm shop and farm machinery; Montana requires journalism and public speaking in the sophomore year; North Dakota wants agricultural geology and entomology; Nebraska, Arizona, New Mexico A. and M., and Texas A. and M. include horticulture, dairy husbandry and poultry during the first two years. The subjects vary. But always the fundamentals for freshmen and sophomores are chemistry, mathematics, English, and basic work in understanding plants and animals.

Ranch boys are sometimes hard to convince that all this foundation work is necessary. "We get two classes of students," says Dr. Howe. "First there are the big old boys who come up with big hats and boots on. All they want is animal husbandry. To hell with English and biology and chemistry. What do they want with those things? Furthermore, when a teacher starts on something they have had experience with, they know it all and give their views quite freely. They are really good kids in most cases. Usually they straighten out in just one semester and stop looking down on the poor instructors who know what is in the books. The other class is less well grounded and maybe less energetic, but they peg along and do fairly well when some of the boys in big hats and boots go back to punching cows, where they belong."

In the junior and senior years the boys can see why they had to take the subjects that seemed like extras. In those years they settle down to a specialty and get ready to use it as a profession. That is when the basic sciences begin to count, and it is just too bad for a student of fertilizers who is not up on his organic chemistry.

There are many pathways to the degree. At Oklahoma A. and M. the course in agriculture involves majors in animal husbandry, dairy, poultry, feed crops, horticulture, agricultural economics, entomology, agricultural engineering, agricultural education, and agricultural journalism. All the big schools offer veterinary medicine. Utah has a separate school of forest, range, and wildlife management. Wyoming even offers special work in running a "guest ranch"—dude ranch, to you—and expects to strengthen its offerings in irrigation.

Estimates vary, but probably not more than half of the graduates go directly into farm or ranch work. Unless a boy has a father already in the business, or can whip up a romance with a ranchman's daughter, or has an independent income, he has two strikes against him as a producer. Other jobs, however, are open to him. He can work for the government, go in for research, teach vocational agriculture, work as a county agent, or sell feed and ranch supplies. He can sometimes become a manager, though that job has its difficulties. Dean McKee at Montana State says he is about through trying to place his graduates as farm and ranch managers with owners who "want a glorified farmhand

259

and try to make all the decisions themselves whether they know anything about the business or not."

For those who go out to spread information, as well as for those who spread fertilizer, there is one final indispensable—practical experience. Ironically, this is the one thing the agricultural college cannot give. The boy must acquire it for himself, and all the school can do is demand that he have it. Montana requires six months on the job before certifying a student for his degree. Texas Tech makes the seniors file an acceptable account of farm or ranch experience before admitting them to candidacy. Deans Blizzard and Willham at Oklahoma A. and M. scrutinize the backgrounds of all their students with a suspicious eye. Many boys with no country experience at all are crowding in, some just shopping and some infected by a romantic concept of life on the range. If they are determined, they are sent out during the summers for close association with hay forks and horses. The deans know where such boys can be placed, but they are very careful about making recommendations. If they send out a dud who washes out about the time the work gets heavy, the college has lost a friend.

Dean McKee thinks the dude-ranch craze causes him some trouble. Boys from the East spend a few weeks riding and fishing, and decide that the rancher's life is the life for them. "We're pretty cold blooded with them," he says. "When a boy lacks experience, we tell him what he's up against."

"Most of our local boys learn," adds George B. Caine, head of the Animal Industry Department at Utah State. "But the Easterners are hard to teach."

The issue of practical training *versus* theory is a major problem in the agricultural colleges today. The question is: How much of what a boy should have learned at home is the college supposed to teach him? Opinion is sharply divided and the result may be a complete revamping of the teaching program.

The man most profoundly disturbed over the question is Sherman Wheeler of Colorado State. "The problem has got to be faced," he asserts, throwing his leathery, loose-jointed frame back in his office chair and waving his pipe. "People are beginning to ask what we are teaching these lads. Will they become better operators, or are we just adding to the personnel of our ever growing bureaucracy? Too many of them go out without the necessary practical knowledge and wind up with a government job at $3,600 a year—more than our instructors get. Farmers and ranchers won't hire them. They are very critical when they pay a boy to do a job, and he doesn't know how.

"Here on this campus we have a division of opinion which will have to be worked out. Some say that a boy doesn't come to college to learn to fix windmills. I say that a boy should be taught to do the things with his hands that he will have to do later. If we don't teach him, he may be driven into a vocational school like California Polytech. Eventually colleges may become upper-crust institutions training men for government or technical positions.

"Meanwhile we are in a bad situation. I have eighty-six students in my class in beef production. There isn't time enough for everybody to try his hand at dehorning, castrating, and so on. This mass-production idea doesn't hit the mark."

At Laramie a mild but single-minded old gentleman plants himself firmly on the other side. "There is plenty to learn in college without learning the trade secrets of the herdsman," Vice President Hill maintains. "That isn't what the students want either. We made a survey not long ago among our graduates. They all said what was needed was more English, journalism, public speaking, and business law. Undergraduates sometimes ask for welding and shop work. They can get those things in the Trade Department of the School of Education, but we think a man big enough to manage an outfit can spend his time a lot more profitably than on welding."

Dean Paul S. Burgess of Arizona is of the same opinion "We don't think college credit should be given for so-called 'practical' courses. These kids need to know about chemistry—nutrition—fungicides. They don't come to college to learn how to run a hay baler."

The tendency seems to be in the direction of emphasis on management and grasp of fundamentals, and away from the trade-school idea. Not many people are on Sherm Wheeler's side all the way, though everybody is conscious of the problem.

"These boys are learning how to be efficient in the production of human food," say the deans at Oklahoma. "It is our job to teach them modern methods. Some fundamental information is necessary, but one can go to extremes in teaching practical subjects—for instance, castrating, docking, and dehorning. Once a boy has performed the operation successfully, he can go on and do it again as many times as he has to. But in agriculture things are changing all the time. The important thing is to give the boy his fundamentals so he can change with them."

Some of the seeds planted by the emphasis on agricultural theory have already begun to sprout. Look, for example, at Texas A. and M., where a heavy research program, work for graduate degrees, and a long yearly schedule of publications channel off a considerable amount of the productive energy of the college. To some of the people of Texas it seems that their proud institution is rising into a rarefied academic atmosphere and leaving the grass roots, from which it sprang, far below. As a result they sometimes send their ranch and farm-minded sons to business school, to law school, or to a liberal arts college for training in economics or business.

Some administrators are just beginning to find out how successfully their specialized offerings are being let alone. Dean W. L. Stangel of Texas Tech tells how it was brought home to him. A rodeo club was organized on the Lubbock campus, and the Dean accepted the sponsorship. To his surprise he found that his best performers were registered in the liberal arts division. When he looked into the matter, he found that many ranch boys, particularly the sons of oil-rich cattlemen from around Midland and Sweetwater, were drifting in, determined to resist chemistry and physics and the advanced theoretical courses to the bitter end. They would take livestock courses, but had not even a languid interest in becoming the well-rounded agriculturists that the college wished them to be.

Dean Stangel feels his responsibility for cultivating even this barren soil. He does not like to think that these boys will go back to their big ranches with no idea of modern agricultural methods —that they will go on ranching as Grandfather did because they don't know any better. "There is going to have to be a new deal here," he declares. "We are not doing what should be done. We are racking our brains now for a suitable curriculum. More work in range management and related subjects may be the solution. We aren't sure yet."

To make sure that the agricultural colleges never get too far from the land and its people, there is the Extension Service. It is through the Extension program that the college goes directly to the farm and ranch. The director of Extension in each state and his staff of county agents and home-demonstration agents are all members of the A. and M. faculty, and the men in the field go back to the campus periodically for conferences which bring them up to date in their specialties. Extension livestock specialists like Tony Fellhauer of Wyoming and Ford Daugherty of Colorado are on the road about two-thirds of the time making demonstrations, holding livestock institutes, grading bulls for livestock pools, setting up range beef production feeding programs, and even undertaking to purchase breeding stock for men who want to improve their herds but don't want to trust their own judgment.

Most effective long-range subdivision of the Extension Service is the Experiment Station. Here the college not merely goes to the rancher—it stays with him. All sorts of tests and experiments are carried on right under the cattleman's nose, and although there was a time when he was not much impressed, he is definitely interested now, particularly if he is young and progressive.

A sample of what goes on may be seen just outside of Dickinson, North Dakota, where a friendly young fellow named Kenneth D. Ford is supervising experimental work on the effect of vitamin A deficiency in range cattle.

Cattlemen know that when the range dries up and stays dry till spring, the cows have difficulty in calving, are short of milk, and produce wobbly offspring troubled with scours and other unfortunate conditions. Lack of vitamin A may be responsible.

Two groups of animals are fed and observed under the worst possible conditions. The cows are bred before going on grass and calve in February, which is all wrong from the cow's point of view. The roughage used is deficient in vitamin A. One group of cattle is fed a carotin supplement to supply the missing vitamins. A second group struggles along as best it can. After three years of experimentation Mr. Ford thinks his results may be trusted.

He considers the problem from the rancher's point of view as well as the experimental scientist's. The carotin capsules are expensive and impractical for ordinary use. Fortified cake might be the answer, and commercial feed companies have it for sale. But cake costs money, and many ranchers want to be sure it is worth the price before they start using it. Kenneth Ford's job is to find out what they want to know, and he expects before long to reach some conclusions.

Dean Walster, who supervises all the experimental work from his office in Fargo, says he has about seventy-five such projects going on.

All sorts of institutions, foundations, and projects call for, and receive, the help of the A. and M. staff. Stop off at Luling, Texas, for a look at the Luling Foundation, and you find yourself in the protecting shadow of Texas A. and M. The Foundation was set up as a public-service institution by millionaire Edgar B. Davis (shoes, rubber, oil) in 1927. Edgar Davis was a pious man, truly grateful for his success and eager to cast some of his bread back upon the waters. He took over 1,223 acres of run-down cotton land, established a trust fund, and proceeded to show the country people how agriculture should be handled. He called on the A. and M. staff to help him. Dean Kyle was on his first board of directors, and a close tie-up has always been maintained with the college. In fact Director Walter Cardwell and General H. Miller Ainsworth, the present board chairman, feel that one of the prime functions of the Foundation is to see that experimental ideas and methods originating at A. and M. are given thorough testing at Luling.

One last function of the A. and M. college is to bring the farmer and rancher to the campus. Several times a year, at most schools, there will be a get-together of some sort. Oklahoma sponsors an annual Livestock Feeders' Day; North Dakota puts on Field Days; Colorado has a Feeders' Day; New Mexico entertains at an Annual Ranch Day. Attendance runs up into the hundreds, and men who would never look at a bulletin will come to listen and learn.

On and off the campus the work goes on with a sort of antlike haste and concentration as students keep coming by the thousands to absorb as much as they can of the agricultural wisdom of the ages and to enjoy that peculiarly American thing we call college life. The Ag Club, the Livestock Club, the Block and Bridle, and a hundred other societies, big and little, give them social contacts, recreation, and friendship. Like students everywhere, they use the college as a marriage bureau where persons of like backgrounds, tastes, and ambitions can get together for connubial purposes. Ranch boys and others of similar interests have even made a fad of rodeo clubs where some of the pattern of life on the range is transferred to the campus.

The formation of the National Intercollegiate Rodeo Association not long ago, it might be added, is likely to put rodeo in the same bracket with football as a leading spectator sport. The effect is already being felt at Wyoming University, where a fairly typical Rodeo Club operates. For the first time in history a member of the Rodeo Club sat on the Student Senate in 1949, says E. K. Faulkner of the Animal Production staff, who acts as sponsor for the group.

The sixty or seventy members include both men and women, and they enjoy each other's society at a large variety of social gatherings—dances, box suppers, and so on. Some of their entertainments are money-raisers, for it costs to send members to competitions out of town. And it should never be forgotten that members are more than rodeo-appreciators. They perform—or try to. And the real point of the whole organization becomes visible only in the arena in front of the cheering stands.

The club has its own grounds with movable bleachers and six chutes—built with money borrowed from the Student Senate. But the real performers do not aspire to be just home talent. They are far travelers and have taken their own horses to Kansas, Colorado, and Texas. A chartered plane took several of them to the first N. I. R. A. rodeo at San Francisco in 1949. Dale Styles came out ahead in riding saddle broncs, but the three ropers, on strange mounts, did not fare so well.

Cowboys and Cattle Kings

Conservative faculty members at Wyoming and other places are not always too much pleased with these clubs, but they are going to have to reconcile themselves to the situation—for rodeo on college campuses seems to have come to stay. And the rules covering eligibility and competition set up by the N. I. R. A. have added standing that the sport badly needed.

When classes and collegiate rodeos are behind them, what happens to ranch boys and girls who graduate from the Ag colleges? Those who go back to the ranch usually take leading roles in the affairs of their home communities. They are looked up to, and more is expected of them. At least they know more about what ought to be done than their fathers did, and most of them are land savers and conservationists instead of exploiters and wasters, as their fathers sometimes were.

They certainly live better. Ideas of a more gracious way of life than their forebears knew come out of the classrooms with them, and the results are visible in the good houses, modern conveniences, and practical arrangements which are the trademark of the progressive rancher all over the country. The lonely ranch house and run-down farm of a generation ago, damned for its ugliness and primitiveness by the Garlands, Cathers, Scarboroughs, and Suckows, is something better now.

Whatever changes for the better have come to the cattle industry in the last twenty-five years must be attributed at least partly to the far-flung, friendly tentacles of the agricultural colleges.

22. Salute Your Honey

"ON BEAR CREEK babies and children always went with their parents to a dance, because nurses were unknown. So little Alfred and Christopher lay there among the wraps, parallel and crosswise with little Taylors, and little Carmodys, and Lees, and all the Bear Creek offspring that was not yet able to skip at large and hamper its indulgent elders in the ballroom.

" 'Why, Lin ain't hyeh yet!' said the Virginian, looking in upon the people. There was Miss Wood, standing up for the quadrille. 'I didn't remember her hair was that pretty,' said he. 'But ain't she a little, little girl!'

"Now she was in truth five feet three; but then he could look away down on the top of her head.

" 'Salute your honey!' called the first fiddler. All partners bowed to each other, and as she turned, Miss Wood saw the man in the doorway. . . .

" 'First lady, centre!' said her partner, reminding her of her turn. 'Have you forgotten how it goes since last time?' "

These goings-on, according to Owen Wister, happened during a barbecue and dance at the Goose Egg Ranch on Bear Creek in western Wyoming at some indefinite time in the eighties. It was a historic occasion for the place and time. There were a barrel of whiskey and a whole barbecued beef out in the yard. There was fiddle music inside for the quadrille, the waltz, and the polka. And there was a whole room full of sleeping babies for the Virginian and Lin McLean to scramble to the bewilderment of their fathers and mothers and to the delight of several generations of readers.

It all happened just about as he said—sometime—somewhere. The Swinton brothers, who gave the barbecue, lived and worked in Wyoming. The Goose Egg Ranch was a famous spread. The scrambling of the babies actually happened some time in the late sixties.[1] Lin McLean and the Virginian and Uncle Hughey and the Taylors and the rest of them were portraits of Wister's Wyoming friends, or composites made out of a number of them.

Anywhere north of the Medicine Bows and south of the Big Horns the descendants of the Carmodys, the Dows, and the Westfalls (whatever their right names were) are still around, enjoying themselves more or less as their grandfathers did. Mr. Ford and Mr. Edison and others have made barbecues and cowboy dances easier to arrange. Nobody has to ride 118 miles on horseback for the sake of a dance with Miss Molly Stark Wood,

---

[1] A story on "A Mixin' of the Babies" was carried by the Dallas *Weekly Herald* for May 4, 1867. It was described as happening "some time ago . . . up North."

as the Virginian did. But the old spirit is there; the fun is just as furious; and they still dance till it is almost time to go back to work the next morning.

Right in the middle of Wister's Wyoming is Sweetwater Hall, where a modern version of the Goose Egg barbecue takes place every two or three weeks, whenever it happens to be handy to have one. The place is seventy miles south of Casper and a short distance from Independence Rock on the Oregon Trail. It stands half a mile from Highway 202 in one of Jim Grieve's pastures on the Dumb Bell Ranch. A half-mile farther down the slope is the Grieve headquarters, and beyond that the Sweetwater pokes about in the bottoms just after breaking through the chasm of the Devil's Gateway. The hall used to be around on the other side of the hill above Tom Sun's place, but it burned down; there was some trouble about the insurance; and it was decided to put the new one up on private property where it could be better looked after.

There is little glamour in Sweetwater Hall. It is a plain frame structure about forty by sixty feet with a kitchen blocked off at one end. The "Him" and "Her" are out behind. No attempt has been made to decorate or even finish the place, and the ceiling is a jungle of wooden cross braces. The floor is plain fir. The only seating arrangements are built-in benches of two-by-twelves the length of each side. A wood stove at each end provides the heat, and five dim bulbs hanging from the rafters cast a dim, but not religious, light. What makes Sweetwater Hall a place of pleasant memories is not the elegance of the appointments but the warmth of the humanity that gathers there—the same warmth that Wister felt.

Tonight it is the turn of the Wayne Sanfords to see that there is a party, which means that Irene has had to do most of the arranging while Wayne punched cows. The notices and announcements have all gone out, and Cactus McCleary is coming down from Casper to provide the music. After supper they take the children over to Grandpa Archie Sanford's house just across the yard and deposit them for the evening. Archie says he and Mrs. Sanford would rather sleep with their grandchildren than go to a dance. Throwing on their party clothes (boots and frontier pants for Wayne, a gray spring suit for Irene), they manage to cover the ten miles to the hall by eight o'clock.

There is plenty to do. The place has not been cleaned out since the last party, and they work from one end to the other in a cloud of dust. By nine o'clock they have boosted the relics of two weeks ago into the starry night and are ready for company.

People are slow in coming. It is necessary to wait; and waiting is not too comfortable, for these May nights are chilly and there is no fire in either stove. The force of habit is strong enough, however, to make Irene Sanford, Mrs. Jim Grieve, and a few others gather round one of them as if it were warm while they swap gossip. Their men fuss in the kitchen getting a blaze started in the range and filling the wash boiler on top with water. Two pounds of coffee in a cotton sack go into the boiler.

Pretty soon Cactus McCleary wanders in, a lanky, red-faced, blond young man. He has brought his accordion, but the rest of the orchestra has been delayed. They are flying down from Casper and have to land eighteen miles away. Somebody will drive them over from the landing field.

Little by little the crowd grows. Mr. and Mrs. Jake Claytor from below Independence Rock come in; the elder Jim Grieves, who live up the road toward Casper; Mrs. Eugene Morehouse, cook at young Jim Grieve's place. Eugene, who doesn't like to dance and says he has a headache, is staying home. His wife is buxom, vigorous, and unable to resist lights and music. Mr. and Mrs. Woodring, formerly of Philadelphia, appear, smiling happily. They came out for a visit and fell in love with the country, just as Owen Wister did. Behind them appear Mr. and Mrs. Billy McIntosh. Billy is a Sweetwater boy who manages for Mrs. Sharp, his wife's grandmother. Virginia has done some teaching in the neighborhood—not at the Bear Creek School, but one very much like it. Billy wears white boots and other emphatic habiliments. He catches the eye but pays little attention to anything but his dancing.

The ranch families drift together on the east side of the hall. A miscellaneous collection of outsiders accumulates on the other side where Cactus is setting up his amplifying equipment. There are three high-school youngsters from Casper, one girl and two boys, in identical green-and-white windbreakers, and people from the oil camp of Bairoil down the road toward Rawlins. The two groups stay more or less separate throughout the evening.

Finally the high-flying members of the orchestra arrive, get their gear arranged, and give signs of being ready. There is a cheerful plunking in the bass and a ripple in the treble from the electric guitar. Cactus leads off with the accordion; the banjo and guitar get on the band wagon; and they settle down together on "Sioux City Sue."

The floor fills at once. Every ranch woman is out there in thirty seconds, and there are no wall flowers. Each man takes care of his own, the married men dancing the first dance with their wives, and the grandfathers stepping to it as vigorously as the youngsters. Ben Grieve takes out Nancy Roper, his steady. Herb Poorman, the shy, good-looking cowboy who works for Archie Sanford, dances with a slender brunette who is obviously fond of him. The oldest of the three Forsberg boys has come over from Bairoil with an ample blonde who looks like Mae West and dances like a young tornado. Forsberg gives away about a hundred pounds and twenty years to some of the kids, but stays well out in front.

Dance follows dance, mostly slow fox trots and waltzes. The high-school crowd gets in a little jitterbugging when "The Wabash Cannonball" comes up, but they subside on the next number, "When My Blue Moon Turns to Gold Again."

A number of male observers lean against the door jambs and drift around in the kitchen. Some are shy, some are strangers, and some have never learned to dance. Half a dozen of them are adolescent boys who are just working up to this sort of thing. They all look on with a fixed intensity, somewhere between pleasure and frustration, at the fun they cannot, or will not, join.

No babies have been parked under the benches as they were at the Goose Egg barbecue, and there is no barrel of whiskey out in the yard. Some of the men have bottles in their cars, however, and little groups stand around outside from time to time and bend an elbow. They drink in turn (owner last) without glasses or chasers.

Tod Jordan looks at Wayne Sanford's contribution and reads

Mary Kidder Rak

J. Evetts Haley

Badger Clark

the label by the light of the just-risen moon. "Well," he remarks, "Wayne Sanford has switched to Calvert."

The ranch women stay in the hall, but seem to take it for granted that the men will do some drinking. Some of the town women put on a rather wet and profane party in one of the parked cars, but this is not part of the pattern for the natives. Yes, they say, people have got tight and started trouble in the past, but it hasn't happened for a long time.

The first part of the dance is not much different from public dances anywhere in the country, but after a ragged rendition of "I'm Sending You a Big Bouquet of Roses" with a faintly rumbling vocal by the guitar player, the atmosphere changes. All of a sudden Don Kortes is up there fiddling with the microphone. Anticipation lights up the ranch families as if a little electric current has passed among them. Don is the best square-dance caller for miles around. Now begins the part of the evening that would make Wister and his Virginian feel right at home.

"All in your places and straighten up your faces!" the microphone booms. "Get your partners for a square dance."

It takes some arranging—"Two couples here" . . . "Three couples over here. . . ."

The musicians get going one by one, as usual, but soon they are hitting it out in a fast rhythm with a heavy beat. The dancers feel it in their muscles. They rock back and forth, flexing their knees. They stamp, clap their hands, and sway their bodies as the pulsing wave of sound penetrates and lifts them. The tune is "Little Brown Jug."

"Form a ring, a great big ring," Don bellows through the microphone. Away they go, five squares, heels coming down on the beat, faces smiling, skirts swirling.

> *First couple out, balance and swing,*
> *Lead right out to the right of the ring.*

If you didn't already know what Don was saying, you would be lost. One middle-aged man in a brown suit, who thought he knew how to do a square dance, looks right and left with a scared expression on his face. Martha Sanford grabs him and hurls him into the figure.

271

Around that lady and peek once more,
Back to the center, and circle four.

Ernie Forsberg comes down hard with his boot heels. Somebody across the hall lets out a shrill whoop. The floor boils with noise and action.

Aleman left.

The man in the brown suit is lost again. The whole whirling square comes apart while he gets himself re-oriented, then plunges back into the pattern. The bewildered visitor is sucked in like a stick being drawn into an eddy of the Sweetwater.

... you know where and I don't care!
That's all!

The noise stops as if somebody had turned off Niagara Falls, leaving a buzzing hush behind as the squares break up and the crowd gathers along the benches.

"Must be time to eat," somebody says—and from under the benches, from the kitchen, from the cars outside, the food boxes begin to appear.

The local arrangement is to have every family bring its own food. Coffee is furnished by the hosts for the evening. There is enough in the big boxes to give everybody more than he can eat, and the providers pass them up and down the benches with hospitable invitations to "help yourself." The bachelors who had no way of making a lunch, the children, the strangers—they all reach in and come back for more.

Pretty little Barbara Jordan, who teaches two children at the ranch school on Archie Sanford's place, has doughnuts, egg-salad sandwiches, and chocolate cake. She passes the box to Jake Claytor, whose own wife and daughter Jean have ample supplies farther down the bench. He "doesn't mind if he does."

While all this is going on, Wayne Sanford and Herb Poorman make a last round to see that each man has kicked in his dollar and a half. They tie a ribbon around his tie or loop it over a button to show that he has paid his debt to Sweetwater society. Then they squat down on their heels by the family food box.

In half an hour the boxes disappear as quickly as they came. Irene Sanford asks Cactus McCleary to give out with some Schottisches and polkas. While he talks it over with his boys, about half the crowd drifts to the door—the old folks, the ones with youngsters, the people who have to drive forty or fifty miles to get home. McCleary leads out on the accordion with "The Beer Barrel Polka," and the rest step to it again.

Two o'clock. Three o'clock. By four o'clock everybody has had enough. There is a final bustle at the door—cheerful good-nights—smothered yawns—startled looks on the faces of people who have just remembered that their day begins at six. No bedrolls out in the yard by a dying barbecue fire this time. Mr. Henry Ford has made that unnecessary. One by one the cars roar up the hill to the highway. Irene and Wayne leave last so they can snap out the lights and lock the door. Another dance is over.

It is four-thirty when they get home. At six Archie, his boys, and his hands will start 350 head of cattle down Sweetwater River toward Pathfinder Dam and summer grazing. Herb Poorman and Ernie Forsberg don't bother to go to bed—just lie down for an hour. Wayne pretends that he is going to get a night's sleep, and actually crawls between the sheets; but in an hour and a half he has on his working clothes and is ready to get behind the herd.

"The Virginian rode away sedately through the autumn sunshine; and as he went he asked his Monte horse a question. 'Do yu' reckon she'll have forgotten you too, you pie-biter?' said he."

23. Corral Branding

SPRING TAKES HER TIME about visiting these Colorado mountain valleys. Although the late April sun is bright and warm and coats come off about the middle of the morning, winter creeps back on little cold paws at night and a man wants a fire and a couple of blankets. At seven thousand feet there is no trifling with the climate.

273

The broad San Luis Valley is all brown grass and gray trees this time of year except for the circle of soaring mountain peaks which fence it in—eastward, the towering ranks of the Sangre de Cristo forever marching south into New Mexico, snow covered from end to end; westward, a lower range, snow spotted and pine furred; far to the north the high white peaks around Mount Ouray; away to the south, the rounded humps of the Black Mountains beyonds the New Mexico line. The level land surrounded by these glittering peaks is wonderfully rich and productive, and even now, when the fields are still asleep, the thick stubble of the meadows, the swollen haystacks and endless irrigation ditches, the clusters of farm and ranch buildings every half mile or so, the sleek Herefords in the pastures—all make wordless promise of riches to come.

Everybody in the valley thinks it is the most fruitful and beautiful region in the whole world, and though the vegetation thins out into sagebrush flats over toward the Sangre de Cristo where the great Trinchera Ranch lies, the rest of the country has plenty of artesian wells and mountain water to make the deep soil go into a frenzy of production every year. In a few weeks, when the awakening comes, the whole valley will be a vivid green with squares of brilliant black where the potato farmers are getting in their crops. The cattle will trail up to the high pastures, leaving the meadows to transform themselves into emerald lakes of grass. And the people will go to work in their little Eden, feeling sorry for the rest of the population which has to put up with inferior portions of the earth's surface.

The cattlemen on the road south out of Monte Vista, in the full swing of the spring branding activities, are out in the pastures every morning as soon as they can see. Mostly they work together. Where the ranches are not too large, here as well as in other parts of the cattle country, neighbors help each other with the big jobs. The atmosphere at these community shindigs is distinctly jolly—not a bit like the serious intensity of group work on the big ranches. Probably the spirit of the old-time husking bees, log rollings, and barn raisings of early America is better preserved at a neighborhood corral branding than anywhere else in the country.

274

Corral Branding

Ted Larick's cattle are to be worked this morning. As Luke McOllough and I clatter down the road in his pickup, we can see that the gang is assembled and the cattle are already penned in the big corral. They must have started driving them in at six—it is eight now and the sun is well up.

The corral is really a battery of corrals. There are three good-sized board pens, a wire pen, and a small board crowding pen which leads into the dipping vat. Nobody uses dipping vats much any more, since spraying techniques have been developed, and the old vat, with trap doors like guillotines at each end controlled by counterweights in nail kegs, looks like the relic it is. Cows and calves, about a hundred of each, are milling and bawling in the biggest board pen and the men are hard at work separating them. The idea is to chase the cows into the wire pen and the calves into another board corral, and if you think it is easy to separate a frantic mother cow from her calf, you try it.

Ted Larick, hawk-faced, wide mouthed, thin lipped, and hipless—a Will James come back to life—is there on a white horse keeping things moving. Bill McNeil from up the road is also on horseback. Bill is a plump, platinum-haired young man with a sharp-nosed, foxy profile, a cheerful disposition, and an unusually flexible and copious profane vocabulary. All the rest are on foot and strung out in a line across the pen reserved for the calves. When a cow and calf charge out of the big corral, impelled by blows and shouts, everybody tries to booger the cow into the wire pen while the calf slips between the men to join its fellows in the far corner of the calf corral. Each man has some sort of persuader, usually a broad piece of leather riveted to a stick, but his main reliance is on running, arm waving, and the most peculiar assortment of shrieks, cries, whistles, yodels, yells, and syllables from unknown tongues ever heard outside an African jungle.

Nearest the gate through which the unwilling cows are being shoved is Homer Winfrey, the oldest man there, a lantern-jawed frontier type with a hat just ready to pass into the heirloom stage, and an incongruous pair of rimless bifocals. He runs and leaps like a jumping-jack to turn the cattle and looks tough enough to keep it up all day in spite of his sixty-odd years. Jack Phillips

275

is next, a decorative young man in a nice, fresh, cream-colored shirt which he would like to keep clean. He doesn't take things quite so hard as Homer does. Luke McOllough, short and cheerful and boiling with energy puts himself and his red-checked shirt into the next position and immediately starts talking things up as if he were a cheer leader at a football game. After a while Luke's son Fritz rides up in full cowboy regalia (except that he is minus spurs), ties his horse to the fence, and gets in on the far end of the line.

It is a noisy, active bunch of men. Everybody hollers, jokes, and cusses according to the impulse of the moment. This is strictly a stag party and the boys let their vocabularies down. Bill McNeil has a greater natural aptitude for cussing than the rest, and seems to have done more to develop his talents. He has a stronger flow and gets more variety. It isn't profane cursing, however, if that makes sense. Nobody stops to savor his own vulgarity or looks around to see if anybody is shocked. The words just pop out like steam through a safety valve or air from a punctured tube, and are about as immoral in their total effect as any other blowout.

Cow work seems to demand a little colorful language. There just couldn't be enough men to make every cow behave. Some break through the line and have happy reunions with their calves. Some of the calves run like terriers and slip through into the cow pen. This is just part of the game and is noticed only by a few "damns" and "hells." When the separation is about completed, the men gang up on the unruly cows and run them by main force into their proper place. Then Ted rides into the cow pen, ropes the misplaced offspring, and drags them out one at a time. The calves are then shooed into the smallest wooden pen in the approach to the dipping vat, and crowded together so close they can't take a deep breath. A gate is swung across to hold them while the branding equipment is set up.

First there has to be a chute. They improvise one out of an old gate and three wooden "panels"—frames which look like gates but will never see a hinge. These are set close together so as to admit one calf at a time from the crowding pen. At the end of the chute the calf table is set up.

Corral Branding

Among these neighbors there is an inventive genius named Shorty Schroeder who doesn't happen to be here today, but has left his indelible imprint on the proceedings. It was Shorty who made the calf table and the furnace for heating the irons, both of them rough and ready, but effective.

The table is a gadget into which the calf squeezes after reaching the end of the narrow chute. It is smaller at the bottom than at the top and is open at both ends. The calf who steps into it thinks he sees liberty just ahead, but he does not reckon with the devlish ingenuity of man. There are levers and springs and pivots on that machine, and just as he reaches the center, a man heaves on a handle which "pinches" the poor little animal by pulling the sides together at the top. Caught in this vise-like grip, the calf is further astonished as the last stage of the operator's heave tilts the whole arrangement, leaving him helpless on his side at the height of a man's waist. What happens to him after that is what happens to any calf at a roundup.

Shorty's work is not the result of original research. Several makes of calf tables are on the market, and all he had to do was to adapt the idea to local conditions. He used heavy iron tubing, roughly but strongly welded, with a few oak timbers here and there. Some of these calves are pretty big and it takes a husky piece of machinery to hold them.

The furnace is the child of his own brain. A section of an old oil drum forms the base, with a piece cut out of one side to let in the butane fuel, and another opposite to put the branding irons through. This is only a third of the thing. A bigger section of oil drum bolts on to the base, and above that the final third tapers into a chimney-like pipe. When set up, the whole assembly looks like an old-fashioned upright steam engine.

It takes a while to put it together and McNeil grows exasperated. "——!" he exclaims when the bricks for holding the irons fall down and he has to take the top of the furnace off so he can start over. He rips out the four-letter word like an enraged laundress tearing a pair of wet overalls in two.

The bricks back in place, somebody remembers that the furnace doesn't draw properly unless the draft is cut off at the joint between the base and the top. "—— it to ——!" McNeil roars, not

277

uncheerfully, "did anybody bring any —— water?" A pailful is produced from somewhere, and he mixes some mud to caulk the crack. "Turn the —— —— thing on!" he yells, holding his muddy hands away from him.

Sudden silence falls. A crescent wrench is needed to turn on the gas, and nobody has brought one. It is necessary to wait while Winfrey goes after one in the pickup.

"—— ——," McNeil remarks, deep in his chest.

During the fifteen-minute wait Ted gets his knife sharp on an oil stone and Luke sets out the needles and the two kinds of vaccine while the talk goes on. The subject is butane explosions from which one or another has barely escaped with his life.

"You remember we tried an upright butane tank first," Luke says reminiscently. "I'll never forget the time a gate blew open and knocked the tank over. I sure thought we were gone. I ran over Shorty Schroeder getting out of there. Since then we've used a horizontal tank."

When Winfrey gets back, they light the furnace and put the irons into the red-hot interior. At last Ted hollers, "Let's go!"

Fritz is in the crowding pen running the calves into the chute. He gets behind the leaders and strains and shoves to pack them in there, nose to tail. A bar across the chute at chest level holds up the front calf until the men are ready to let him into the squeeze.

Phillips is assigned to the calf table, somewhat to his dismay. They show him where to take hold and how to apply his weight. He asks some questions, tries a couple of practice pulls, and says he is ready. The first calf is prodded into the machine. It goes into action faster than Phillips expects and is halfway out at the other end before he gets into his heave. When the table tilts, the poor animal hangs like a shirt tail half way out of a man's pants, bawling pitifully. Phillips lets the table down again and the calf comes out like a cork. Luke flanks him and they work him over on the ground in the traditional manner.

Phillips is a little better with the next one, and on number three his timing is perfect. As the calf reaches dead center, he swings back and down on the lever, and the contraption responds, squeaking and banging.

By —— —— ——!" yells McNeil triumphantly, "You're moving in fast now!"

Ted handles the knife, castrating and earmarking. Winfrey works on the front end, slipping his needles into the fold of brisket or neck. Luke and McNeil slide back and forth like machines on well oiled rails between the furnace and the table, putting cold irons in the fire and taking out hot ones. Carefully they apply the 5 on the left hip, taking the iron off for an instant when the calf kicks and struggles, pressing it down again on the smoking hide when he quiets down, touching up a skimped edge to make sure the impression is clear. Phillips has a straight bar which he applies to form the slash of the Five Slash brand.[1]

It soon falls into a sort of rhythm. The men work hard but not furiously. There are minutes when everybody can breathe briefly and maybe light a cigarette or pass a remark to his neighbor. It is a good sample of a cattleman's morning this time of year, full of vigorous action, reminiscences of other times, and commentaries with the bark on. To watch such a branding session is to understand why cattlemen like the life in spite of the blood, dirt, and sweat. Whatever else may be said about their job, it takes a man to do it.

The cows bawl and surge in the wire pen. The calves bleat and squall as the needle and the knife bite briefly into their hides. The sun makes good headway into the warmth of mid-morning. "... they caught that Mexican two years ago and he got eighteen months—had the hide and some of the beef hid in his wood pile."

"Yeah, he's out now and trying to find something else to steal."

"The —— —— —— don't know how to operate," observes McNeil. "Luke and I know how, though. I eat Luke's beef and he eats mine so we come out even."

They all enjoy a companionable guffaw. "You wait," Luke tells him. "The beef is really going to get eaten when they put up the irrigation project back there in the hills. Those ditch diggers will get fat on local meat."

[1] *5/*

McNeil looks at his watch. "Well, —— —— me for a —— ——. We aren't moving too fast."

"Wait till we get used to this thing," Ted says. "This is only the second year we've used it."

"It's a lot easier on the calves," Luke adds. "Come on boys, let's get back at it. Twenty-five to go and we can eat."

"—— —— —— ——; —— ——," McNeil bellows, bending over the furnace. "What are we waiting for?"

T 24. Pen, Pencil, and Paint

HE CATTLEMAN never has been very good at putting himself on paper. Although he is the most photographed, most written about, and most persistently painted of all human objects, the men who photograph, describe, and draw him are seldom cattle people themselves. For one Eugene Manlove Rhodes, for one Charley Russell, for one Will James, even—there will be a dozen Emerson Houghs, Stewart Edward Whites, and Frederic Remingtons who get close to the heart of the business but are not really a part of it. Owen Wister was a tenderfoot. Ross Santee had graduated from the Chicago Art Institute when he came to Arizona to begin the serious business of his life; and now that he is famous, he has gone East to live. Conrad Richter (*The Sea of Grass*) was a curious Middle Westerner who spent several years in New Mexico taking copious notes and writing excellent prose, and then went back where he came from. Walter Van Tilburg Clark (*The Oxbow Incident, Track of the Cat*) migrated to Nevada when his father became president of the University of Nevada, and he has served his time as a teacher himself. J. Frank Dobie (*The Longhorns*) was a ranch boy and never has got over it, but he, too, left the corral for the classroom, and when he feels like branding something now, he goes out and burns the hide of his garage door.

Cowboys and cattle raisers are men of action, not men of words. Few and far between are the ones with an urge for self-expression. Rudolph Mellard (*Hills and Horizons*) and S. Omar

Barker (*Buckaroo Ballads*) are among the few who can belong to the life and interpret it, too. It is probably highly significant that the commentator who appeals to the widest audience of cattlemen in the northern states is a Sheridan, Wyoming, newspaperman named H. F. Sinclair who writes under the name of Neckyoke Jones. He reports imaginary conversations with his pal Greasewood for eight periodicals, mostly stockmen's magazines, in a style somewhere between Bill Nye and Will Rogers, giving the government hell and making common-sense observations on current events.

"I see where our Congrissman has come to git busy with the post bar an' the wire stretcher to do a little fence mendin'," I sez to Greasewood, as we was drivin' over the chuck holes in the county road early this week. "Yep," sez Greasewood, who runs a few head of cows along with me here on Long Pine Crick, an' who is awful smart, "you know Congrissmen ain't like us stockmen. We got to patch fence up about twict a year—but a Congrissman don't do no fence repairin' exceptin' at long intervals. . . .

"Yessir," sez Greasewood, "an' he is allus tryin' to guess which way the dill pickel will squirt. The city folks wants low prices—an' the farmer, he wants high prices—labor wants to ramrod the outfit—an' industry is spurrin' the boys where the hair is short. I would guess that a Congrissman is a most unhappy feller."[1]

Neckyoke has done some ranching, talks the ranchman's lingo, and expresses the ranchman's ideas. Everyone likes him, and I have seen him at a cattleman's convention, a stocky man with a bush of gray hair and a happy, rosy face, surrounded by ranch people who wanted his autograph. But he does not make his living handling steers. The men who speak for or interpret the American cattle raiser seldom do. Most of them are enthusiastic immigrants who discover what he has always known, and tell the world about it with a convert's passion and pride.

This may not be such a bad thing in the long run. Interpreters who live too close to their subjects seldom see those subjects in focus and are apt to suffer from a sort of artistic myopia. Some of our most prominent regionalists have been infected. The cul-

[1] *The Montana Stockgrower*, November 15, 1949, 35.

ture, the customs, the folklore, the history of their native range fill the entire stage for them. As the cattlemen say, "Their fence line is the end of the world for them."

When one of these enthusiasts gains a hearing, the first thing you know he has put on priestly robes and is finding mystical meanings in all sorts of ordinary things. In the Southwest the mesquite is fast becoming a sacred plant. The javelina and the longhorn look more and more like the unicorn and the gryphon. Even the unheroic coyote, thanks to Mr. Dobie, may now be looked upon as a spiritual being with more intelligence than many of his human brothers. And every old cowpoke is in danger of being bracketed with Jeremiah as a prophet and seer.

Under ordinary circumstances, of course, the old cowpoke can no more see himself in perspective than he can take observations of the back of his own neck. On the other hand almost any honest observer can report essential facts, whether he belongs to the tribe or not. Does an editor refuse to send a sports writer to describe a prizefight unless the man has cauliflower ears?

This may help to explain why the most significant interpretive work on the cattleman seems to come from men and women who are not firmly rooted in the cow business. Sometimes, like E. E. Dale of the University of Oklahoma, they gain breadth of vision by going from ranch life into something else. More often they enter the cowman's world from another background, acquire a good thick coating of range lore, and then go on to their real job of interpretation.

Take Badger Clark as an example of the way it often works. He is poet laureate of South Dakota and a maker of rhymes about the cowboy which even the cowboys like to read, but he was born in Iowa and spent his youth on a prairie farm at Deadwood. In 1906 a spot on one of his lungs, the after-effect of a fever picked up in Cuba, sent him to Arizona. He was twenty-three years old when he got his first taste of life on the range on the old Cross I Half Circle twelve miles north of Tombstone. He loved it right away.

For four years he was a contented cowboy for the Kendall brothers. His feelings about the new life were too potent for prose, and he began to put his letters home into verse. His mother, with fine business sense, sent "Ridin'" off to the *Pacific Monthly*, and immediately—just like that—the editor forwarded a check for ten dollars.

"That ruined me," says Badger Clark. "I've been doing it ever since."

He has published two volumes of verse (*Sun and Saddle Leather* and *Grass Grown Trails*) based mostly on his Arizona experience, and one book of tales titled *Spike*. He likes poems better than stories because, in his opinion, "everybody reads stories and nobody remembers them. Of course the financial return is better, but although only about one reader in forty goes in for verse, the ones who do are nuts about it, quote it, discuss it, and sometimes write letters to the author."

Badger's verse is conventional in its rhythms and not subtle, intellectual, or spiritual in its quality. But it has a sort of epitomizing effect which makes it live in the minds of many people, especially if they have strong fundamental feelings and longings with limited power of expression. Badger says for them what they want to say and can't. As a result there is a timelessness about his work which keeps his best poems, like "The Cowboy's Prayer," from becoming dated. It was written during his first summer in Arizona, but is going yet. And every now and then he gets a letter from somebody who has just seen it and thinks it is something new. Practically everybody who reads has come across a line or two from "The Old Cow Man":

> *Let cattle rub my tombstone down*
> *And coyotes mourn their kin,*
> *Let hawses paw and tromp the moun'*
> *But don't you fence it in!*

Remarkable also is "The Westerner," written forty years and more ago, which says what most cattlemen are now thinking about the sleight-of-hand performances going on in Washington:

> *I dream no dreams of a nurse-maid state*
> *That will spoon me out my food.*
> *A stout heart sings in the fray with fate*
> *And the shock and sweat are good.*

283

People wonder about Badger when they first see him. He dresses to suit himself—cavalry boots, riding breeches, military blouse, khaki shirt, black Windsor tie. He has a gentle, humorous, goateed face where wrinkles, kindliness, and whiskers struggle for first place. His smile lights him up like one of those old-fashioned lamps with a translucent china shade. There is no affectation in him, in spite of his odd costume. Just individuality.

The boots and poetic tie have their practical uses, too. Badger is a great reader of his own poems, and his external oddities are good attention catchers when he gets up in front of an audience. After the ice is broken, he needs no help from anything. He plays his pieces for every ounce of feeling there is in them, but never runs over into too much sentiment. He acts just a little, but not too much; reads just enough, and quits.

A contented bachelor, he lives in a cabin in the Black Hills where eight or ten tame deer can come to his door every day. He has pleasant companionships also with the bluebirds who nest on his porch and the pack rats who steal the beads off his moccasins.

At the other end of the cattleman's world lives John Joseph Mathews, historian of the Osage Indians of eastern Oklahoma. Since the Osages live in the midst of some of the richest grazing lands in the country and raise a great many cows themselves, Joe Mathews has added his portion to the annals of cattle land.

He, too, is on the outside of the industry, looking in. His father, William Shirley Mathews, was a grandson of Old Bill Williams, the mountain man. As a member of the Osage tribe he moved into eastern Oklahoma and started a cattle ranch on Beaver Creek when his people migrated from Kansas in 1872. This home place is still owned and operated by members of the family. Son Joe never lived on the ranch, but he spent much of his boyhood there riding and working about the pastures.

The important turn in the road came when he went away to school. "About the time I was a sophomore at the University of Oklahoma," he says, "a member of the staff, a former Rhodes scholar, noticed me, and told me I was the type who would get something out of going to Oxford. He made all the arrangements for my entrance and pushed them through. I never had thought of such a thing myself."

He took his B. A. Oxon. in Natural Science in two years and read modern history for two more. He believes this English experience was responsible for his passionate interest in the history and lore of his own region. "It gave me a perspective—built up my enthusiasm," he thinks. "It showed me what I had." Further insight into his life and his future job came with travel in Europe and Africa. In 1930 he came back and started talking to the old men, going through material accumulated by his father, and visiting historical collections in other states. Since *Wah-Kon-Tah* (1932) and *Sundown* (1934) he has been the unofficial interpreter of his people to the rest of the world and spends much of his time in Washington working with the Office of Indian Affairs.

There is little about Joe Mathews to suggest his percentage of Indian blood, except perhaps his eyes, which are a little remote and weary. He is a stocky six-footer with big shoulders, but his total effect is one of elegance rather than energy. He is a cultivated gentleman with a well-modulated voice and a fine flow of conversation about everything under the sun, but especially about the business of writing history and reading books. Nevertheless one part of him belongs to "those wild horsemen with eagle feathers spinning in their roaches, and shirts ballooning in the wind as they seemed to flow down the hillsides and across the ravines"—the Osages who in days gone by held up the herds from Texas shouting, "Beef! Beef! Beef!"

Joe Mathews has done what nobody else could do in showing the human hearts beating under the blankets of those stately red men, but he is man of many interests and has marked out a dozen other trails through the history of his region. His biggest difficulty at the moment is deciding what to do next. He has just finished a life of the late E. W. Marland, founder and long-time president of the Marland Oil Company and once governor of Oklahoma, and is ready for another job. "But," he says, "I have so much stuff stacked up I don't know which way to turn."

The ranch women who have anything to say about themselves are not run-of-the-mill, either. Agnes Morley Cleaveland

(*No Life for a Lady*) grew up on a New Mexico ranch but she took a degree at Stanford and lived for years in Berkeley, California. Her friend Mary Kidder Rak never even lived on a ranch until she was a married woman. She was another Stanford graduate who had worked at teaching, social service, and writing in California. Then she met Charley Rak, a Texan from the Devil's River country who was studying forestry at the University of California. They married and eventually went to live on ranch property in the mountains of southeastern Arizona. One of their places was called Hell's Hip Pocket. After 1919, Mary Rak was a confirmed cattle woman.

Ten years went by, and she and Charley had a visit from Dane Coolidge, well-known writer of Western stories. In the lobby of a Douglas hotel he introduced her to Harrison Leussler of Houghton Mifflin Company. "Mrs. Rak writes, too," he said.

"Do you?" Leussler asked.

"Oh, no, I can't sell what I write, so I make hooked rugs and give them away."

He pulled out a card. "If you do write something that might interest us, send it to me."

Coolidge nagged at her about writing a ranch book. "I'm too busy doing ranch work," she told him. Then Charley started nagging. Finally she gave in.

"I'll write one chapter and send it to Leussler," she promised.

She did it, and Leussler was pleased. The result was *A Cowman's Wife* (1934) and *Mountain Cattle* (1936). Of all the books which attempt to tell about life on a modern cattle ranch, these two seem to me the truest and best. Mrs. Rak has learned, without losing her sense of humor, that out there in the hills "if it isn't one thing it is another," and that ranch women always "must play second fiddle to a cow."

After two such powerful efforts she feels that she has exhausted her ranch material "for the moment," and her output now is mostly magazine articles.

I first saw her at the door of her dobe house in Rucker Canyon high up in the Chiricahuas and a long way from town. The place is floored with broad pine planks—all she and Charley could get during wartime when they built it. The walls are plain

ARTIST AT WORK
Harold Bugbee finds a temporary clearing
in his studio

white plaster. Next to Charley, Mrs. Rak seems to love Navajo rugs and dogs. She has a fine collection of the latter in a constant scramble to get in and out of the house. They are remarkably vital and vigorous dogs, and they have to be to keep up with as vital and vigorous a person as their mistress.

She is a solidly built woman with white hair and a bearing of great dignity—a Republican, an Episcopalian, a respecter of order and decency with no compromise in her. She dresses like a ranch woman in cotton dresses and comfortable shoes, but the house is full of books and the Underwood portable on the card table in her big bedroom gets lots of exercise. She could never see the inside of her business so clearly if she had not brought so much in from the outside.

The formula seems to be about the same for an artist, and we have the foremost living interpreter of the cowboy in line and color to prove it—Harold Bugbee of Clarendon, Texas.

Harold was born in Massachusetts, and still has a trace of the Pilgrim Fathers in his speech, particularly in his vanishing "r's." In the early days his uncle, T. S. Bugbee, had one of the biggest ranches in the Panhandle. Clarendon was then in the middle of one of his pastures, though it was moved over to the railroad later on. For years T. S. had been trying to get Harold's father to come on down to Texas, but it was not until 1913, when Harold was thirteen years old, that they finally made the move. However, Harold had been roping chickens and otherwise preparing for cowboy life since he was knee high to a duck.

The cattle range has been his study and delight ever since. He never made his living as a puncher, but he served his time on a horse, came to know the cattle people intimately, and became a better Westerner than most of the natives.

After two years in college he decided that he was taking a roundabout way to realize his ambition, which was to transfer the life of the range country to a piece of drawing paper. He journeyed over to Taos to persuade Herbert Dunton to accept him as a pupil. Dunton, however, was not taking on any students and suggested that he go up to Des Moines, Iowa, to see what he could learn from Charles Cummings.

Cummings was then the big man in the Art Department at

the University of Iowa and at the same time head of a school of his own. Harold stayed with him for two years. At the end of that time Cummings told the boy to go on home. "I have taught you to use your tools," he said. "Most students need another two years to find out what they want to paint, but you already know."

Harold came back to Clarendon and started making sketches, hundreds and thousands of them. He says he probably made more than Charley Russell did. He haunted the Goodnight ranch where there was a herd of buffalo. At certain times of the year he could ride in among them and sketch by the hour, and the buffalo would pay him no mind at all. The country was full of horses, and he worked on them. He went on camping trips with Dunton and visited places where he could draw wild animals. Now he is not afraid to tackle any Western subject, and his great pride is the fact that the men who know are always impressed by his accuracy of detail. Mrs. Bugbee, a tall, lively brunette, thinks his best quality is his versatility. "He can do anything and do it well," she says. "Some artists develop a sort of sameness—particularly when they work with models in a studio. But nobody could look at one of Harold's paintings and say, 'That's a Bugbee sky.'"

Bugbee himself is a mild-mannered, husky, friendly fellow without pretensions and without unnecessary humility. He has a good head of curly black hair, wears boots and riding breeches, and has a notable collection of red plaid shirts. He lives in a small white house on the edge of Clarendon, but does his work in an unbelievably cluttered studio in the basement of his father's old ranch house two or three miles away. He also milks cows and takes care of thirty head of stock.

Mostly he works at black and white illustrations now, but he has done a great deal of painting and has specimens all over the house. He likes to take them out and display them—the deer in the forest, the horses in the red afterglow of sunset, the grizzly bear in front of tremendous gray rocks, the herd of white mustangs that appeared on the prairie about 1883 and went away to nobody knows where. He puts them in a frame one by one to show them to the best advantage. His favorite, the one he thinks is just right, is a band of Indians coming over a hill in the dusk. "It has quality," he thinks. Mrs. Bugbee believes it has "the whole poetic sweep of the Indian people in it."

Some of Bugbee's best work has been done in collaboration with his long-time friend J. Evetts Haley, who lives over at Canyon, fifty miles from Clarendon, when he isn't working cattle on his small ranch above Stinnett or dashing off in pursuit of material for another book. The two of them form one of the most interesting and authentic writing and illustrating teams in the country—certainly the best in the cattle business. A Haley book illustrated by Bugbee is about as flawless as such a thing can be. For if Bugbee is the foremost artist working with cattle people, Haley is the foremost biographer, and one of the best living Western historians.

Haley is the great exception to all that has already been said. Nine times out of ten the man with the pencil may be a refugee from the range, or an immigrant to it; but then along comes a Haley to keep us from saying "Always!" He is a practicing cattleman who can and does interpret his own people.

His first breath was a healthy sniff of Texas air inhaled on a farm in Bell County near Austin. While he was still a youngster, the family moved out to ranch country near Midland and settled on the Concho River. Later they lived in Midland, where the elder Haley was sheriff for a while. In 1921 the boy entered West Texas State College at Canyon and became addicted to history. After taking his M.A. degree at the University of Texas, he came back to a job with the Panhandle-Plains Historical Society at Canyon. He was supposed to visit around among the old-timers getting interviews, letters, newspapers, business records, pictures, museum pieces, and anything else of historical value. Evetts knew how to spit and whittle with the old cowboys. He was a success.

In 1928, after two years of this, he was offered a similar job with the Texas State Historical Association with headquarters in Austin. For eight years he covered the Southwest, digging up historical treasures in unexpected places. Then politics got him.

Evetts has always been what might be called a radical conservative. He is suspicious of panaceas, dislikes waste, and hates bureaucracy. He cannot stand governmental paternalism and despises a system in which the moral fiber of the people is weakened by dependence on doles, subsidies, allotments, controlled prices, and disguised payments for votes. Furthermore, he regards it as a right, a privilege, and a duty to let the world know where he stands on these matters. His outspokenness made it advisable for him to quit his job at the University and has involved him in countless arguments since.

One time Evetts and Harold Bugbee were on their way to San Antonio. They stopped for gas at San Marcos and noticed a gang of WPA workers ostensibly laboring across the street. The efforts of this gang were so feeble and so far between that Evetts could not stand it. He went across and gave those men a real, old-time cussing out.

"I thought we were going to have to fight our way out, or else run," says Bugbee. "But the men were so startled they didn't do a thing, and we drove off."

After severing his connection with the Historical Association, Evetts became a ranch manager. He worked for Dent and Wrather of Dallas on a spread in the Painted Desert of Arizona, and for J. M. West of Houston in West Texas. In 1940 he bought a small ranch for himself in the breaks of the Canadian in the Texas Panhandle, and owns another in partnership with his brother southwest of Midland. He trades in cattle, pastures steers, and does a considerable amount of ranch business in spite of the increasing pressure of his historical work.

He bears the tokens of a ranchman's life, including a choice collection of broken bones and impressive areas of scar tissue. The latest episode took place in the spring of 1949 in New Mexico when he was helping to work some snuffy steers in a corral. One old outlaw came along the fence. "I boogered at him, but he didn't booger," Evetts says. The critter ran right over him, and so did the rest of the herd, most of them stepping on his face. They broke a rib, tore another one loose from its moorings, cracked a bone in his neck, left his lower lip hanging halfway to his chest, and banged him up all over. "They like to killed me," he summarizes.

The thin scar on the left side of his face and some good story telling about the wild ride to Albuquerque in search of a doctor are the only relics of that experience. His wind-burned, still-boyish face and lean cowpuncher's frame are the same as ever.

The first book he published, and in some ways still the most notable, was *The XIT Ranch of Texas* (1929), a history of the enormous ranch empire, stretching two hundred miles north and south along the New Mexico line in the Panhandle, which was turned over to the Capitol Land and Cattle Company in payment for the construction of the present state Capitol. The heirs of the Farwell brothers, who were the leading spirits in the company, commissioned him to do the job.

He did it thoroughly and frankly, as he does everything. With the help of Ira Aten, an old-time Texas Ranger who successfully policed the ranch and stopped the rustlers from cleaning the company out, he named names and stated facts with the chips falling in every direction. The result was a suit for one million dollars directed at Haley, Aten, and the Farwell heirs. There was a trial at Lubbock, full of high feelings and potential gunplay.

It is said, though Evetts refuses to discuss the matter, that he went for a ride one afternoon and was left afoot in the wilds along the Canadian River, thereby almost missing his own trial. Nobody knew that country better than he did, however, and he got back in time.

The first suit was won by the defendants, but there were many more coming up—enough to keep juries busy acquitting Haley for years. The Farwells settled out of court for "a considerable sum"—nobody will say how much. Haley's book was withdrawn from circulation, and only part of the edition ever came on the market. Copies sell for as much as $75 now, when they can be found. Scarce as it is, however, no bibliography on the cattle country is complete without it.

The XIT business set the pattern for the rest of Haley's career. He has stayed close to the pasture ever since. His material and his attitudes are fairly accurately compressed into the first paragraph of the introduction to *Charles Goodnight*:

This book is more than the biography of a man—it is the background of my own soil, a part of my own tradition. Every wind that drifts the alkali dust from the Goodnight Trail across my home range suggests a land of cattle and horses; every damp breeze carries the penetrating fragrance of greasewood, suggestive of the bold life that rode along it. Yet the land and its life have hardly changed with the years. Today, our trails are still the trails of cattle; our problems those of aridity, of grass and water along the bitter Pecos. . . .

The biographical side of Western history has called forth Haley's best efforts. The deepest chords in his nature respond to indomitable, many-sided, salty human beings—what he calls "broad-gauged" personalities. He has done notable books on Goodnight, the pioneer cattleman; on Major Littlefield, cowman and philanthropist; on the Schreiner family, ranchmen and merchants of Kerrville; and on Jeff Milton, cavalier gunfighter and peace officer. At the moment he is digging up history about the sheep, cattle, and oil region around San Angelo.

Haley has traveled hundreds of thousands of miles in pursuit of material, and has accumulated an enormous stock of information—all methodically filed away in a battery of steel cabinets in the study of his Spanish-style home in Canyon. Along with the enthusiasm which drives him over every trail in the Southwest, he has the meticulous conscientiousness of the professional researcher and checks, cross-checks, and rechecks every detail. He writes in a straightforward, unself-conscious style with enough color to carry his facts to the general reader. The combination of accuracy, completeness, and readability which he has developed make him, in the eyes of many competent judges, the foremost historian of the cattle country, and one of the best in the land.

He is the outstanding exception to the rule that the cattleman has to hire somebody else to tell his story.

T 25. I'm Leavin' Cheyenne

HIS BOOK is not finished, and never could be. It would take a lifetime to do it right—to cover all the ground, see everybody who should be seen, and know even a small part of what should be known. And by that time everything would have changed so much that it would be necessary to start all over. There would always be wilder country farther on, new ranges to explore, better stories to tell. I leave my cattleman friends with great regret. I wish I had met the rest of them.

I wish I could have gone to Florida and seen the results of four hundred years of ranching in a state which nobody but a Floridian would call cattle country. I wish I could have called on Channing Cope in Georgia and seen what he has done to rebuild the Georgia soil and turn it over to fat cattle—driven through Mississippi where the ruined cotton fields are changing to grass lands—seen the great ranches in Washington and Oregon—visited the gigantic feed lots in California.

I wish I had seen Bill Spidel at Roundup, Montana, and called on W. S. Tash at Dillon before he died. I would still like to have an hour with Charley Redd far out in the Utah desert and another with Monte Ritchie, the Britisher who operates the JA Ranch in Texas. Some day I want to talk with Turk and Alice Greenough of the great family of rodeo performers at Red Lodge on the edge of Yellowstone Park, and accept Leo Cremer's invitation to come and see him at the ranch near Big Timber where he raises rodeo stock for shows all over the country. Next time I cross the Nebraska Sandhills I want to stop and see Sam McKelvie, publisher, former governor, and venerable cattleman, and on the way up I would turn aside to Manhattan, Kansas, to hear Dan Casement talk. If I could, I would follow the crooked trails that lead to the doorway of Rafael Carrasco at the mouth of Fresno Creek in the wilds of the Big Bend country of Texas, and I would get acquainted with Bryon Conley, who operates a saloon in Havre, Montana. The cowboys say that you can blow in all you have on Saturday getting drunk in his bar, sober up on Sunday, and go to work at his ranch on Monday at the prevailing wage—for one week exactly, so you can have a little to move on with.

I wish I could try again to catch Simpson Walker at home in his battleship-gray frame house beyond the Cimarron south of Freedom, Oklahoma, and hunt up Don Short on the Maltese Cross where Theodore Roosevelt kept the irons hot while better cowman did the branding. I might have more luck next time meeting Robert Yellowtail at his farm on the Crow Reservation (his daughter Mrs. Laura He-Does-It was the only one at home when I called), and I would like to go back to Rapid City for one more chance at former Governor Tom Berry. I want to ask him if it is really true that he is the only man in the country who could put his feet on Roosevelt's desk in Washington and talk to him as one cowman to another.

But what I wish most is that I could travel farther down some of the dim trails, winding across the prairies or climbing into the hills, which lead to those far-away little ranch worlds lost in the desert or buried deep in mountain folds. I would ask nothing better than to load up with extra oil, water, and gas and head east across the Utah desert from Torrey to see what Robber's Roost is really like. I should like once more to come down that long and lonesome road in western Wyoming which wanders south out of Jackson's Hole and follows the emerald thread of the Green River with the Wind River Mountains, snowy and remote, slanting off to the southeast. I would stop at Big Piney (population 242) on Saturday night and see what sort of business the beer parlor along the highway does, and maybe go to church the next day to find out if any cowboys showed up at both places.

And if there is any way I can manage it, I want to see Boulder, Utah.

From 1879 to 1932 Boulder was the home of three hundred Mormon souls who lived a happy life completely cut off from the rest of the world. The place was settled when Brigham Young's scouts combed every corner of Utah for the smallest habitable nook where one of the saints could settle. They found this fruitful spot on the east slope of Utah's central backbone of mountains looking out over the tremendous waste of stone and sagebrush which stretches clear to the Western Slope of the Colorado Rockies. A number of Mormon fathers—John King, Beeson Lewis, Amasa Lyman, Levi Brinkerhoff, William G. Meeks—brought their families and formed a community which might as well have been a colony on another planet.

Until the CCC built a road in during the early thirties, the business of the town was conducted by horse power. Cream was packed out on muleback, and supplies were packed in the same way. Once a prosperous citizen wanted a Model T Ford, and he got one by taking it to pieces and loading the parts on pack horses. The cart track that eventually opened up over the rugged slopes to Escalante was so steep in spots that footholds had to be chiseled in the rock so the horses could make the climb. When winter set in, of course, there was no further intercourse with the outside world until after the spring thaw. Three hundred men, women, and children did not seem to mind.

The road from Escalante is passable now in all weathers, but the trail which comes in from the north is not so good, and those who travel it after a rain run the risk of sliding a couple of thousand feet down a mountainside and spending the rest of their lives at the bottom of a canyon. The heavens opened the day I tried to get in. I took the advice of experts and turned back.

Sometimes the citizens think a little wistfully of all the good tourist dollars which could come their way from near-by Bryce Canyon National Park. "If the mountain roads were made safe they would quickly become a regular route for tourists, since travelers agree that for rugged, varied, and colorful scenery the Boulder country is unsurpassed," wrote Nethella Griffin twelve or fourteen years ago in her manuscript history of the town. The roads have been improved since then, but there are still very few tourists in Boulder.

I am glad of it. I am glad that there are still places where people who stick to the highways and look for good hotels will never go. I like to see those uncluttered white spaces on the map traversed by one thin meandering line of unimproved road which leads to an "x" called Palomas Ranch or Twin Windmills. At the end of that lonesome road I know there is an indestructible American who explains cheerfully to his infrequent visitors: "All we've got here is a lot of brush and some mighty good neighbors."

I am glad that man is there to keep alive some of the things this country has always stood for. I hope he lives forever in the untouched and untouchable eternity of his plains and mountains —that he may keep his brush and his good neighbors with the good will of his fellow citizens till hell freezes over and Gabriel blows his horn.

Not From Books

BUT FROM PEOPLE comes most of the material in these pages. Documentary sources are acknowledged in footnotes, and many of the men and women who helped are quoted under their own names and in their own words in the text. I owe much to a number of others who provided the background for chapters which had to be left out for lack of space, and even more to a thousand casually encountered souls whom I have seen or watched or spoken to without always finding out their names. My heartiest thanks and deepest gratitude are due particularly to the people listed below, whether their opinions and experiences appear in the preceding pages or not:

ARIZONA

Ash Fork: Mrs. Francis W. Campbell
Douglas: Mrs. Dean Bloomquist; Mrs. Mary Kidder Rak
Paulden: Mr. and Mrs. Adams of the Wishing Well
Phoenix: R. L. Cole, credit officer, Office of Indian Affairs; Rich Johnson, associate editor of the *Arizona Farmer;* Charles M. Morgan; Mrs. Dixie Stewart and Mrs. Elsie Haverty of the Live Stock Sanitary Board
Prescott: A. L. Favour
Tombstone: Mrs. Ethel Robinson Macia
Tucson: Dean Paul S. Burgess of the State University; Mrs. Mary B. Price, director of the State Lunch Program; Florence Reese, rural supervisor
Williams: L. R. Lessel, supervisor of the Williams Division, Kaibab National Forest; Mrs. Merle Cowan and Fay Udine of the Williams Chamber of Commerce

Not from Books

Castle Rock: Josef Winkler

Cheraw: Henry A. Bledsoe

Craig: G. N. Winder; Earl Van Tassel

Crestone: Mr. and Mrs. Alfred M. Collins

Delta: Floyd Beach; Editor H. R. Holliday; 4-H Club Agent Everett Hogan; County Agent David Rice (now secretary of the Colorado Stock Growers and Feeders Association)

Denver: F. E. Mollin, executive secretary, and Radford S. Hall, assistant executive secretary, of the American National Livestock Association; D. O. Appleton, editor of *The American Cattle Producer;* Herbert O. Brayer, former state archivist; John T. Caine III, general manager of the National Western Stock Show; Arthur H. Carhart; Martha Daiss of the Colorado Dude Ranchers Association; Mrs. Alys Freeze of the Denver Public Library; Forest Service specialists Don Bloch and H. E. Schwan; Matador Ranch employees Miss E. A. Hall and Sam Wiley; Willard Simms, editor of *The Record-Stockman;* Agnes Wright Spring; Fred Rosenstock; Marvin Russell, editor of the *Colorado Rancher and Farmer*

Fort Collins: S. S. Wheeler, head of Animal Husbandry; Eugene B. Bertone, W. E. Connell, and H. H. Stonaker of the Animal Husbandry Department; Ford Daugherty, Barry Duff, and Cecil Staver of the Extension Division—all on the staff of Colorado A. and M. College

Greeley: Warren Montfort; Mrs. Harvey Witwer and Mrs. Stowe Witwer

Gunnison: Dr. Lois Borland; Clyde and Jim Buffington; Dr. George Nuckolls

Golden: George S. Green

Hayden: Isadore Bolten; Mr. and Mrs. Farrington Carpenter; Mrs. Lou Fulton

Kremmling: Mr. and Mrs. Fred C. DeBerard

La Junta: County Superintendent Edna Campbell

Lamar: Mr. Spannagel of the Soil Conservation Service; County Superintendent Nan S. Creaghe; Mrs. Raymond McMillin

Meeker: A. J. Bloomfield of the AAA office; Undersheriff Earl J. Stewart

Monte Vista: Ted Larick; Mr. and Mrs. Lucas McOllough; Mrs. Orville Worley

Steamboat Springs: Forest Ranger D. E. Gibson; C. W. Light; Sheriff William McFarlane; County Superintendent Vivian Maxwell; Bill Ross

Whitewater: C. V. Hallenbeck

KANSAS

Liberal: County Agent V. S. Crippen; Henry Hatch; Mr. and Mrs. A. H. Keating; Jimmy Keating; Mr. and Mrs. Bernard Limmert; Louis Limmert

Syracuse: County Agent William S. Bort

MONTANA

Billings: M. A. Johnson, director of extension, Indian Service; Walter C. Nye, secretary of the Dude Ranchers Association; D. V. Ross; W. E. Rawlings, acting regional director, Bureau of Reclamation; Marcus Snyder

Big Timber: Wallis Huidekoper

Bozeman: Dean Clyde McKee of Montana State College; 4-H Club Leader T. W. Thompson

Crow Agency: O. W. Davidson, farm management supervisor; Mrs. Laura He-Does-It; Superintendent L. C. Lippert; E. M. McAulay, government auditor; Range Manager G. I. Powers

Forsythe: Mrs. Grace G. Edmiston

Glendive: J. F. Langendorff of the SCS; Chester Nentwig of the sheriff's office; County Superintendent Opha Suckow

Havre: Smoky Downing

Helena: Ralph Miracle, secretary and executive officer of the Montana Livestock Commission; E. A. Phillips, secretary of the Montana Stockgrowers Association

Rosebud: Mr. and Mrs. Valley Kowis; Mrs. Bill Hayes

Shawmut: Leo J. Cremer

Wyola: Lawrence Fuller

NEBRASKA

Alliance: Chase Feagins, chief brand inspector and secretary of the Nebraska Brand Committee; W. A. Johnson, secretary of the Nebraska Stock Growers Association

Chadron: County Agent Harry Kuska; Mr. and Mrs. Hudson D. Mead; David Rice, work unit conservationist; Walter and Wilford Scott; Joe Webster

Hyannis: J. H. Monahan and Earl Monahan

Not from Books

Lincoln: J. J. Deuser, soil conservation supervisor

Mitchell: Dr. C. R. Watson, president of the Nebraska Stock Growers Association

Omaha: Harry Coffee, president of the Union Stock Yards

New Mexico

Anthony: Mr. and Mrs. Harry B. Morris

Albuquerque: Wiley Brown; Horace Hening, secretary of the New Mexico Cattle Growers Association

Clovis: Dr. and Mrs. Ross Calvin; Mr. and Mrs. Wesley Quinn

Magdalena: G. W. Evans, former president of the New Mexico Stock Growers Association

Marquez: Lee S. Evans

Mesilla: Columbus McNatt

Silver City: Mr. and Mrs. Huling Means

Willard: Mr. and Mrs. Lloyd Brandenburg

North Dakota

Alexander: Anders Madsen

Beach: Don L. Short

Denbigh: John C. Eaton

Dickinson: Kenneth D. Ford, associate animal husbandman, Experiment Station; Mr. and Mrs. T. G. Saunders

Elbowoods: R. N. Christiansen, range supervisor for the Indian Service

Fargo: M. L. Buchanan, head of the Department of Animal Husbandry; Ernest De Alton, head of the Department of Agricultural Education; Ruth Shepard, associate state 4-H-Club leader; Dean H. L. Walster—all of North Dakota State Agricultural College

Keene: Angus Kennedy

Killdeer: Leo D. Harris; Sam Rhoades

Minot: Odd A. Osteroos, secretary of the North Dakota Stockmen's Association

Watford City: Mrs. E. C. Henderson; Mr. and Mrs. Lyle Henderson; Frank Keogh; A. K. Kennedy, Jr.

Oklahoma

Alva: County Agent Marvin Wright

Burbank: Bob Donelson

Buffalo: Bill Bland

Norman: W. S. Campbell, Ellsworth Collings, and E. E. Dale—all of the University of Oklahoma

299

Stillwater: Dean W. L. Blizzard; W. R. Felton, assistant state superintendent of vocational education; J. P. Perkey, director of vocational education; Dean O. S. Willham—all of Oklahoma A. and M. College

Oklahoma City: Governor Roy Turner

Pawhuska: Claude Higgins; Claude Huideburg; John Joseph Mathews; Dee Martin; Sam Rogers; Creed Utt

Ponca City: Mrs. Charles McNeese; Sam Steele

Sulphur: John Blenkin; Dr. Glynden Easley; Summers Hudson; Mr. and Mrs. Jim McClelland

SOUTH DAKOTA

Brookings: W. F. Kumlien, professor of rural sociology, South Dakota State College

Custer: Badger Clark

Cheyenne River Agency: L. E. Holloway, agricultural extension agent

Lower Brule Agency: Paul L. Howard, range manager

Midland: Tom Jones; Thurman Adkins

Piedmont: Mrs. Ernest Ham; Wayne Keith

Pierre: Bill Ames and Lois Hipple of the Pierre *Daily Capitol Journal;* Ralph Vandercook

Rapid City: Sheriff Earl Gensler; S. D. Ham, district club agent; Henry P. Holzman, associate animal husbandman; J. E. Horgan, president of the South Dakota Stockgrowers Association; W. M. Rasmussen, secretary of the Association

TEXAS

Albany: Wat Mathews; Joe E. Reynolds; County Agent W. C. Vines

Alpine: Joe Pardoll

Altair: Mr. and Mrs. Lester Bunge

Amarillo: Mrs. P. K. Brian

Austin: Miss Winnie Allen, Dr. H. Bailey Carroll, Ruth Hunnicutt, and Frances Oliver—all of the University of Texas; Mrs. Marian Stoner

Bastrop: John Barton

Beeville: Jim Ballard; Editor Camp Ezell; John Impson; A. C. Jones

Brownsville: Joe Canales; Judge Harbert Davenport; Clarence Laroche

Bryan: John Ashton; Extension Editor Louis Franke; Everett C. Martin, Tad Moses, and J. D. Pruitt of the Extension Division; Ide

Not from Books

P. Trotter, former head of the Extension Division—all of Texas
A. and M. College

Canyon: Mr. and Mrs. J. Evetts Haley; Boone McClure, curator of
the Panhandle-Plains Historical Museum; L. F. Sheffey of West
Texas State College

Catarina: Bessie Pearce; the Reverend and Mrs. J. Stuart Pearce

Clarendon: Mr. and Mrs. Harold D. Bugbee; Mr. and Mrs. Jim Mc-
Murtrie; 'Pinky" Price; L. T. Shelton

Columbus: Mrs. Henry Frnka; Superintendent Marley Giddens;
Miss Lillian Reese; Herbert Schroeder; Elbert Tait; Tanner
Walker

Corpus Christi: Charles G. English

Cuero: Henry Blackwell; Chief Deputy C. J. Durant; County Agent
Joe M. Glover; Mrs. C. T. Traylor

Dallas: Wayne Gard; W. S. Henson; Jack Langston, Sr.; Mrs. Olive
Swope

Del Río: Richard Hawkins; F. C. Whitehead

Dickens: County Superintendent Robert Williams

Elgin: Judge and Mrs. John L. Danly; Mr. and Mrs. Ed Fromme;
Judge and Mrs. C. W. Webb

El Paso: Tom Burchell; Mrs. Tom Crum; Joe Evans; Carl Hertzog;
Ralph Holmes; Mrs. J. V. McAdoo; Dace Myres; Buck Pyle; Mrs.
O. L. Shipman

Floydada: Mrs. Lucille Allred; Judge John Chapman

Fort Davis: Worth Evans

Fort Stockton: Mr. and Mrs. Farris Baker; Mr. and Mrs. Arthur Har-
ral; Mr. and Mrs. Gloster Harral; Mr. and Mrs. H. H. Matthews;
Mrs. Grace Moore Martin, county home demonstration agent;
Genell Slaten

Fort Worth: Henry Bell, secretary of the Texas and Southwestern
Cattleraisers Association; Henry Biederman, editor of *The Cattleman*

Georgetown: Mrs. John F. Yearwood and Miss Eunice Yearwood

Guthrie: Coy Drennan; George Humphreys

Hempstead: William Compton; Mrs. Matt Crook; Miss Barbara
Groce

Houston: John S. Kuykendall; Bill Tipton

Iraan: Superintendent C. B. Downing

Kingsville: J. W. Howe and R. J. Cook of Texas A. and I. College;
Mr. and Mrs. Richard M. Kleberg; Dr. J. K. Northway

Linn: Mr. and Mrs. Argyle McAllen

301

Lubbock: Dean W. L. Stangel of Texas Tech
Luling: General H. Miller Ainsworth; Director Walter Cardwell and John Love of the Luling Institute
Marathon: Mrs. Margaret Buttrill; Guy Combs
Marfa: Mrs. L. C. Brite; Miss Patty McKenzie; Mr. and Mrs. Rudolph Mellard; Judge O. H. Metcalfe; Mr. and Mrs. Clay Mitchell
Matador: Henry H. Schweitzer; Manager Johnny Stevens
Meridian: H. J. Cureton, Jr.; Mrs. Ruby Nichols Cutbirth; Claude Everett; Craig Logan; Mr. and Mrs. Stillman Nichols; Bruce Parks; County Attorney Sam Smith; County Superintendent Joe White
Ozona: Mrs. Jane Augustine Young
Premont: Alvino Canales; Gus Canales; Rafael and Sara García; Charles Hornsby
Port Lavaca: W. L. and C. S. Traylor
Rawls: Bill Giles
Río Grande City: Mrs. Florence Scott
Roma: Joe Guerra; Virgil Lott
San Antonio: C. Stanley Banks; Col. M. L. Crimmins
San Angelo: Editor Houston Hart; M. D. Fanning, manager of the Chamber of Commerce
Schulenburg: Mrs. Joe Lessing
Seminole: Mrs. Harriet Rothé
Seymour: Charley Cocanower
Stamford: Judge Charles S. Coombes; Editor John M. Dewees; Mr. and Mrs. John Selmon; A. J. Swenson; Mr. and Mrs. A. M. G. Swenson; W. L. Wilson, manager of the Chamber of Commerce
Smithville: Adolf Adamcik
Uvalde: County Agent D. P. Gallman; Miss Mary Hoag; Superintendent M. B. Morris; Hugo Mika, of the SCS; the Reverend H. Addison Woestemeyer
Valentine: Mr. and Mrs. R. N. Everett, Sr.; Mrs. Alfred Means
Victoria: Henry Koontz
Walnut Springs: Mr. and Mrs. Elmer Campbell; Gene Harwell; S. M. Martin; Bill Roberts; Albert Rogers
Weatherford: Mr. and Mrs. J. Y. Crum; Key Curl and John Horn of the SCS
Wellington: County Superintendent B. W. Beaird; James A. Leach
Wichita Falls: County Agent Max Carpenter and Assistant County Agent Bill Pallmeyer; Lester Jones; Dr. and Mrs. Austin Leach; Bob McFall; T. J. Waggoner, Jr.; Merle Waggoner

Not from Books

Bicknell: R. J. Brinkerhoff
Boulder: Nethella Griffin
Heber: L. C. Montgomery, president of the Utah Cattle and Horse Growers Association
Kanab: Merlin Shumway
Logan: Dr. George B. Caine, head of Animal Industry at Utah State Agricultural College; Morris Taylor, livestock marketing specialist
Salt Lake City: Judge Martin M. Larson

WYOMING

Alcova: Mr. and Mrs. Jim Grieve; Mr. and Mrs. Archie Sanford; Mr. and Mrs. Wayne Sanford; Tom Sun
Banner: Miss Jennie Williams and Mrs. Nona Williams
Casper: Lon M. Claytor; Mr. and Mrs. George Snodgrass
Cheyenne: Bob Lazear; Charles Thompson of the *Wyoming State Tribune;* Russell Thorp, former secretary of the Wyoming Stock Growers Association; Fred E. and Francis Warren
Cody: Fred Wright
Douglas: County Agent W. L. Chapman; H. P. Pollard; Mr. and Mrs. Jack Thompson
Encampment: Andy Anderson
Gillette: Mr. and Mrs. Ray H. O'Connor
Jackson: Felix Buckenroth; Pete Hansen; Mr. and Mrs. Cliff Hansen; Mr. and Mrs. Wallace Hiatt; Archie Pendergast, secretary of the Chamber of Commerce
Laramie: Harold W. Benn, administrative assistant to the Dean of Agriculture; Albert W. Bowman, director of extension; E. K. Faulkner, assistant professor of Animal Production; Vice President John A. Hill, formerly dean of the College of Agriculture; Burton Marston, 4-H specialist; Peggy Varvandakis, assistant editor, Agricultural Extension Service—all of the University of Wyoming; Oda Mason; Oliver Wallis
Lusk: G. S. Mill; County Agent B. H. Trierweiler
Sheridan: Mrs. Floyd Reno
Rawlins: County Superintendent Helen Irving; County Agent Melvin Lynch

OTHER STATES

Samuel R. Guard, editor and publisher of *The Breeder's Gazette*, Louisville, Kentucky

Ladd Haystead, Wallkill, New York

Franklin M. Reck, Detroit, Michigan

Alan Rogers, Ellensburg, Washington

Mrs. Frances Clausewitz, Miss Ruth Drees, Mrs. Mary Virginia Olson, Mrs. Nita Qualtrough, and Miss Sarah Jane Stockwell, who helped with the manuscript

Index

Index

307

Index

Index

311

Index

313

Index

UNIVERSITY OF OKLAHOMA PRESS

NORMAN